SILVER & STONE

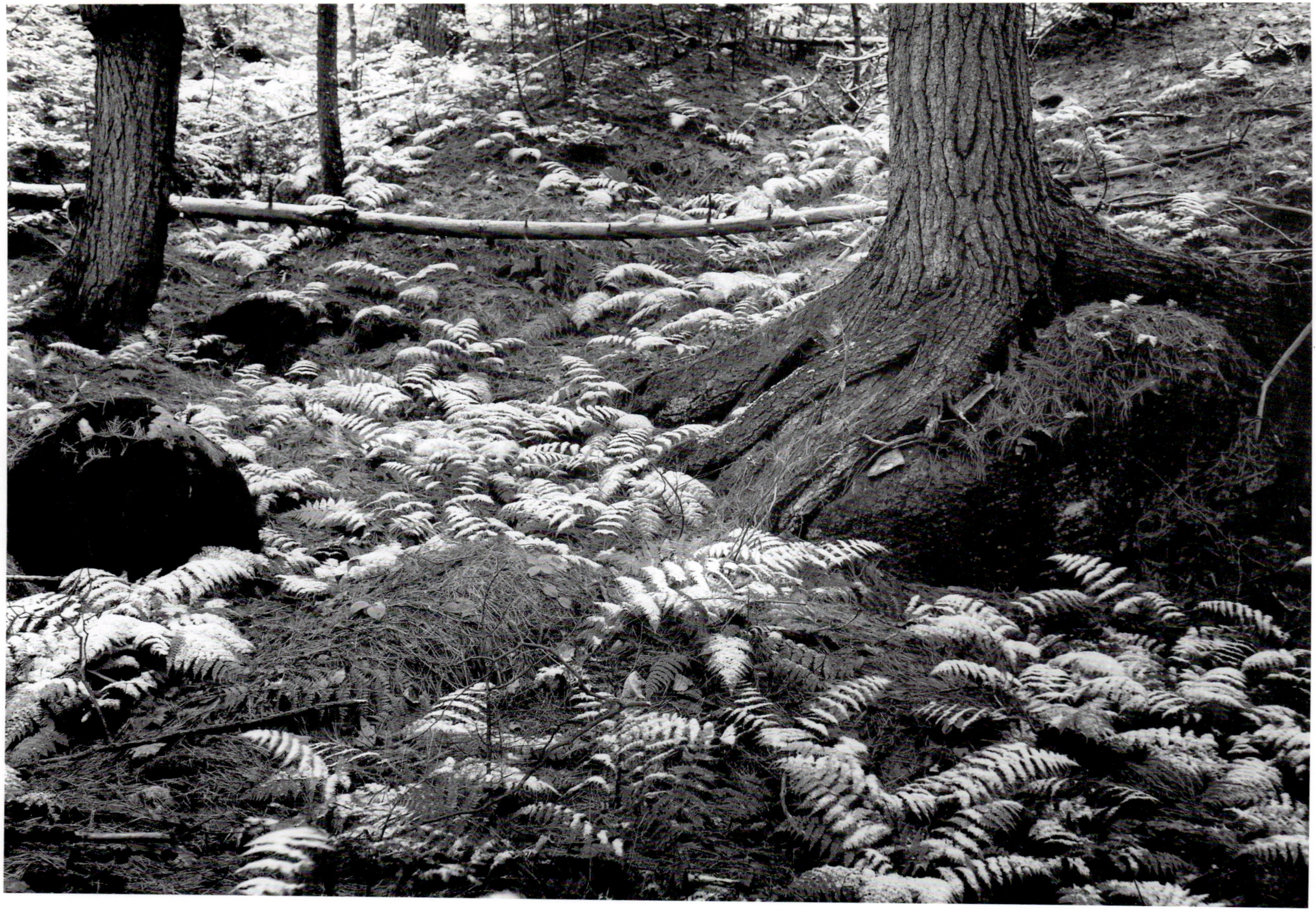

FERNS AND SNOW NEAR MUSKOKA RIVER 1994

SILVER & STONE

LYLE McINTYRE

Stoddart

A BOSTON MILLS PRESS BOOK

Canadian Cataloguing in Publication Data

McIntyre, Lyle, 1937-
Silver and Stone

ISBN 1-55046-114-1

1. Georgian Bay (Ont. : Bay) – Pictorial works.
2. Muskoka (Ont.) – Pictorial works. I. Title

First published in 1994 by
Stoddart Publishing Co. Limited
34 Lesmill Road
Toronto, Ontario
M3B 2T6
(416)445-3333

A BOSTON MILLS PRESS BOOK
The Boston Mills Press
132 Main Street
Erin, Ontario
N0B 1T0

Design by Gillian Stead
Printed in Hong Kong

The publisher gratefully acknowledges the support of the Canada Council,
Ministry of Culture, Tourism and Recreation, Ontario Arts Council and
Ontario Publishing Centre in the development of writing and publishing in Canada.

FOR JANICE

Contents

FOREWORD

I am honoured to have been asked by Lyle McIntyre to write a brief foreword to his book, *Silver & Stone*. This beautiful book features a series of photographs taken by Lyle along the northeastern shore of Georgian Bay and in Muskoka, a region that I know intimately, having grown up there. Lyle has managed to capture the true essence of this region in these remarkable photographs.

Northeastern Georgian Bay and Muskoka comprise a beautiful and unique part of the world, one that possesses many moods and many lights. The basis for all photography is light and its use, and I have examined each of these photographs from the perspective of seeking knowledge as to how light has pervaded each subject. In extraordinary fashion, Lyle has been able to capture the region's moods of light and thus make these photographs transcendent.

It takes infinite patience, quickness of eye, and the adroit use of one's hands to transform the camera from a mere tool into a means of great artistic expression. Lyle's artistic ability and scientific approach have allowed him to preserve these magical moments of natural topography and to stir our innate love for nature in its purest forms.

The Honourable John B. Aird

PREFACE

The important relationship of silver in black-and-white photography, and the mixture of rock and stone that makes up the Muskoka and Georgian Bay landscape, led me to title this book *Silver & Stone*.

Photography is my vehicle, the ground glass of my camera the window to another world, one far more beautiful and exciting than the one I know from day to day. The world seen through my camera is without conflict; it is quiet and peaceful and found only when I look for it. Photography allows me to sense and feel without being disturbed, to drink in nature, and to escape the everyday boundaries created by forces I do not know or understand.

The camera can not only record the images before it, it can also evoke deep emotions. And while colour photography can record a certain beauty, black-and-white photography can communicate texture, mood and depth to a greater degree. This was one of the reasons I chose to work in black and white.

The images in this book were chosen because of my pure love of the land—its four seasons, its lakes, its wetlands, its shorelines. I hope that, through these photographs you will be able to feel what I felt while searching out and photographing these special corners. Most of the locations I photographed are not mentioned on maps or other documents, but all are found in the unique and wonderful region of Muskoka and Georgian Bay.

Lyle McIntyre
Bracebridge, Ontario
June 1994

THE PLATES

WINDSWEPT PINE AND STONE, GEORGIAN BAY 1993

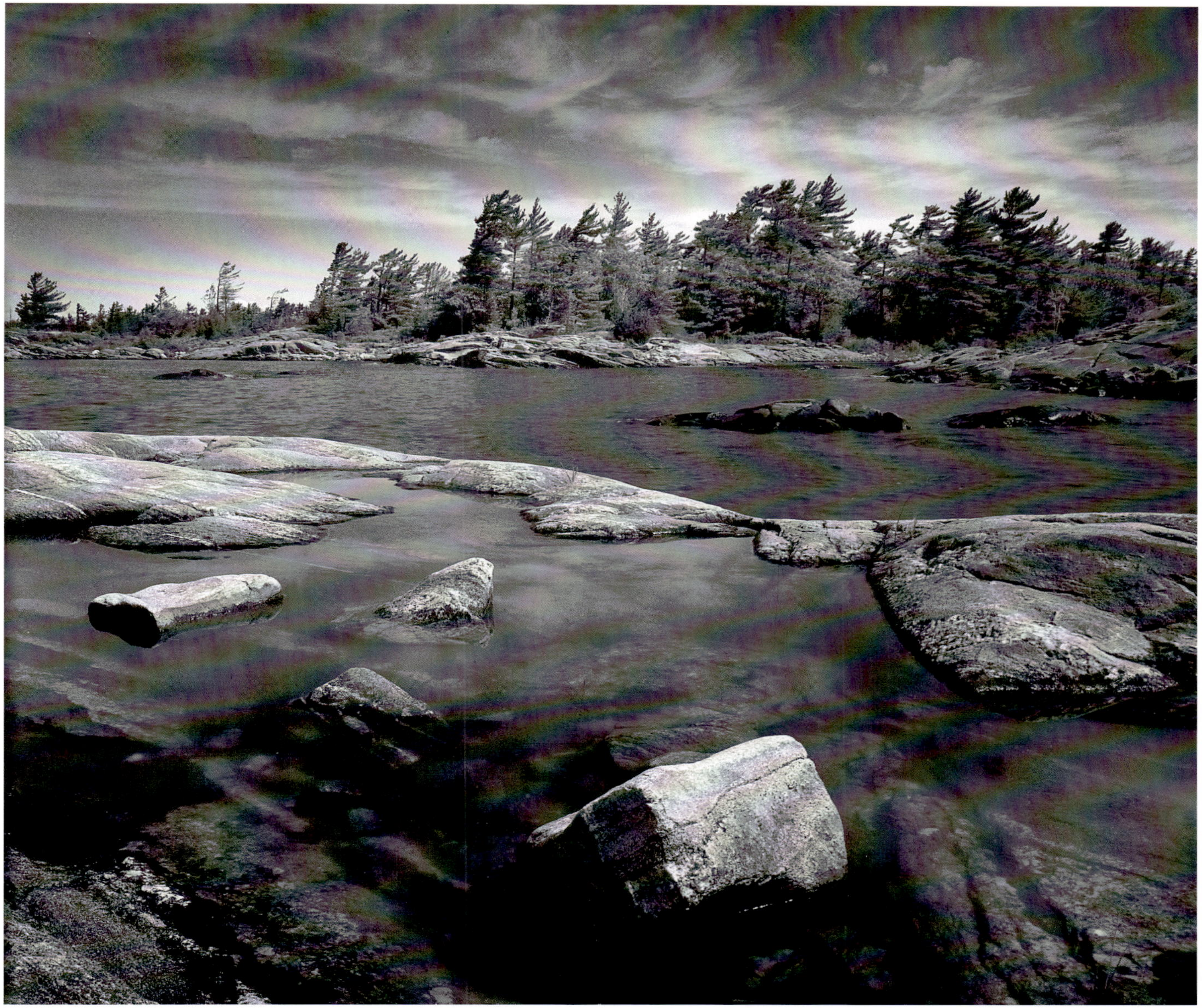

PLATE 1

ROCK AND LIGHT, TWELVE MILE BAY 1993

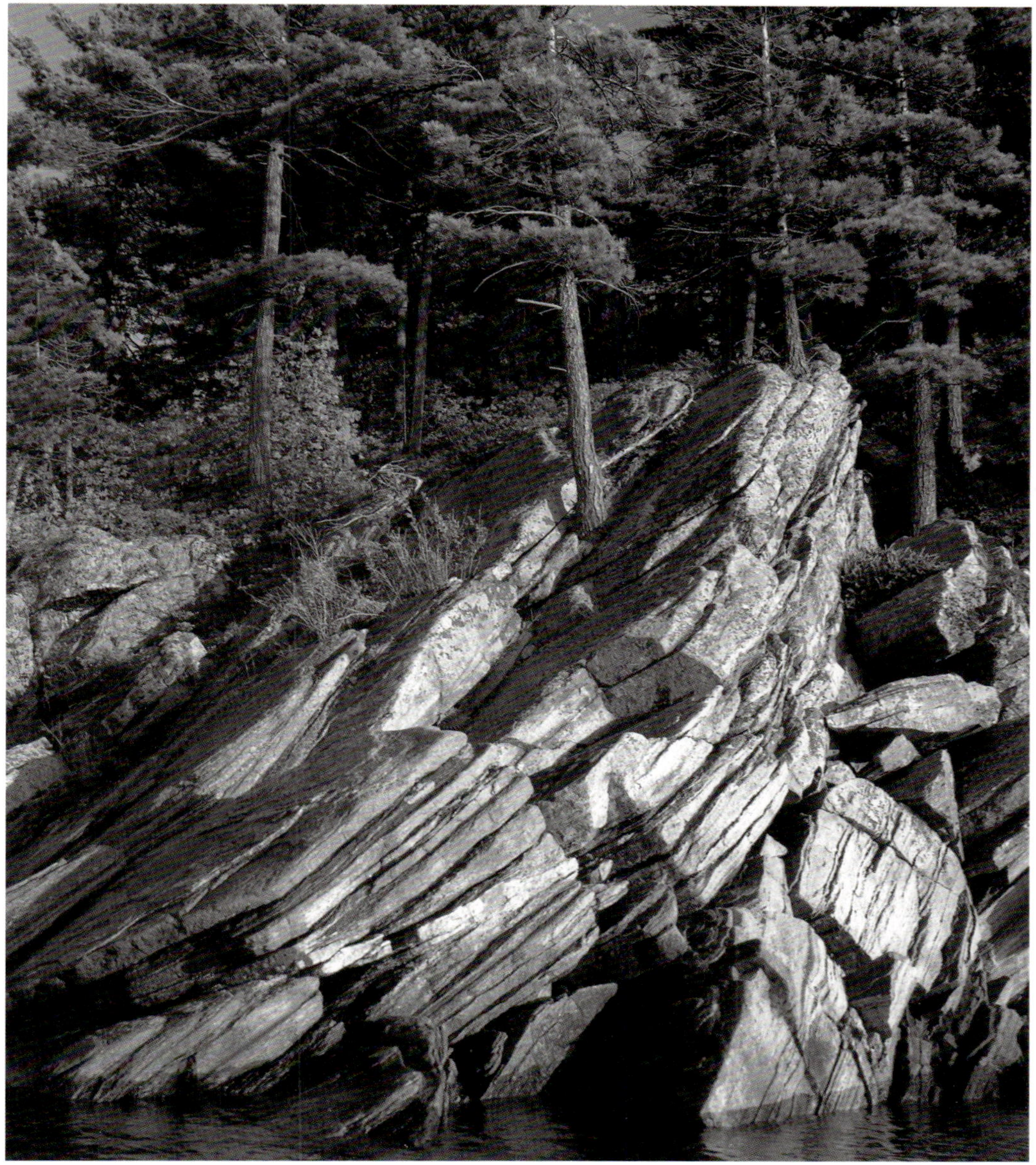

PLATE 2

STONE SHAPES, WRECK ISLAND 1992

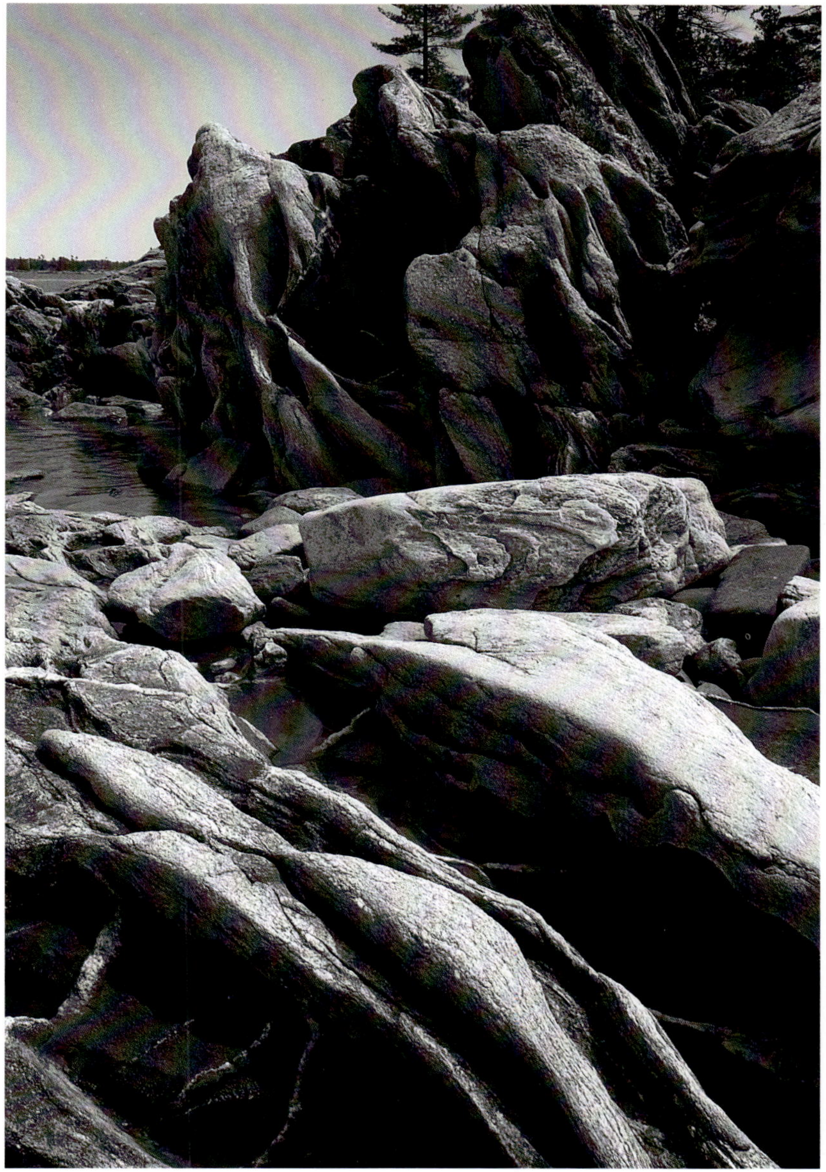

PLATE 3

WATER AND STONE, WRECK ISLAND 1993

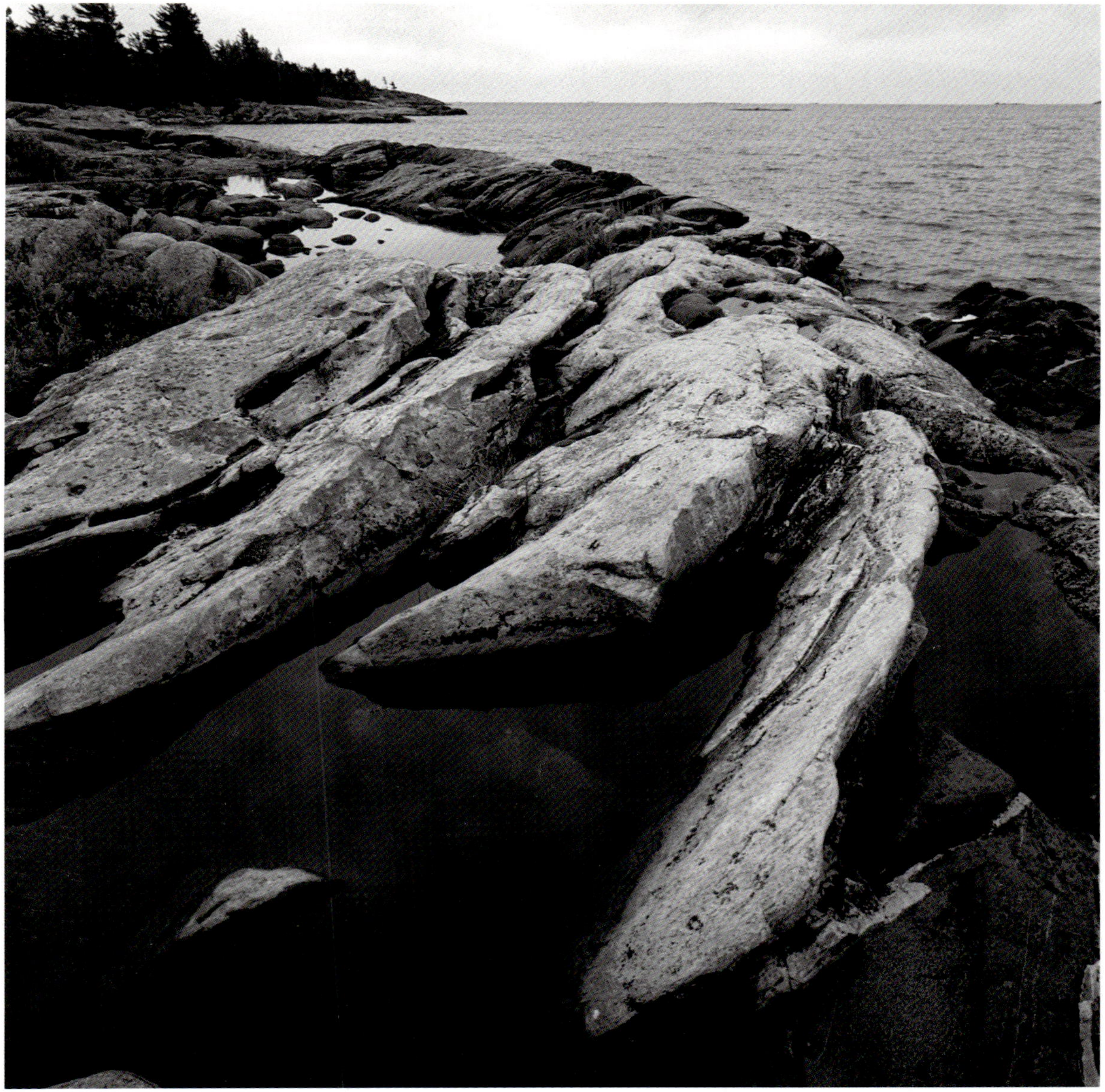

PLATE 4

SURF AND STONE, GEORGIAN BAY 1993

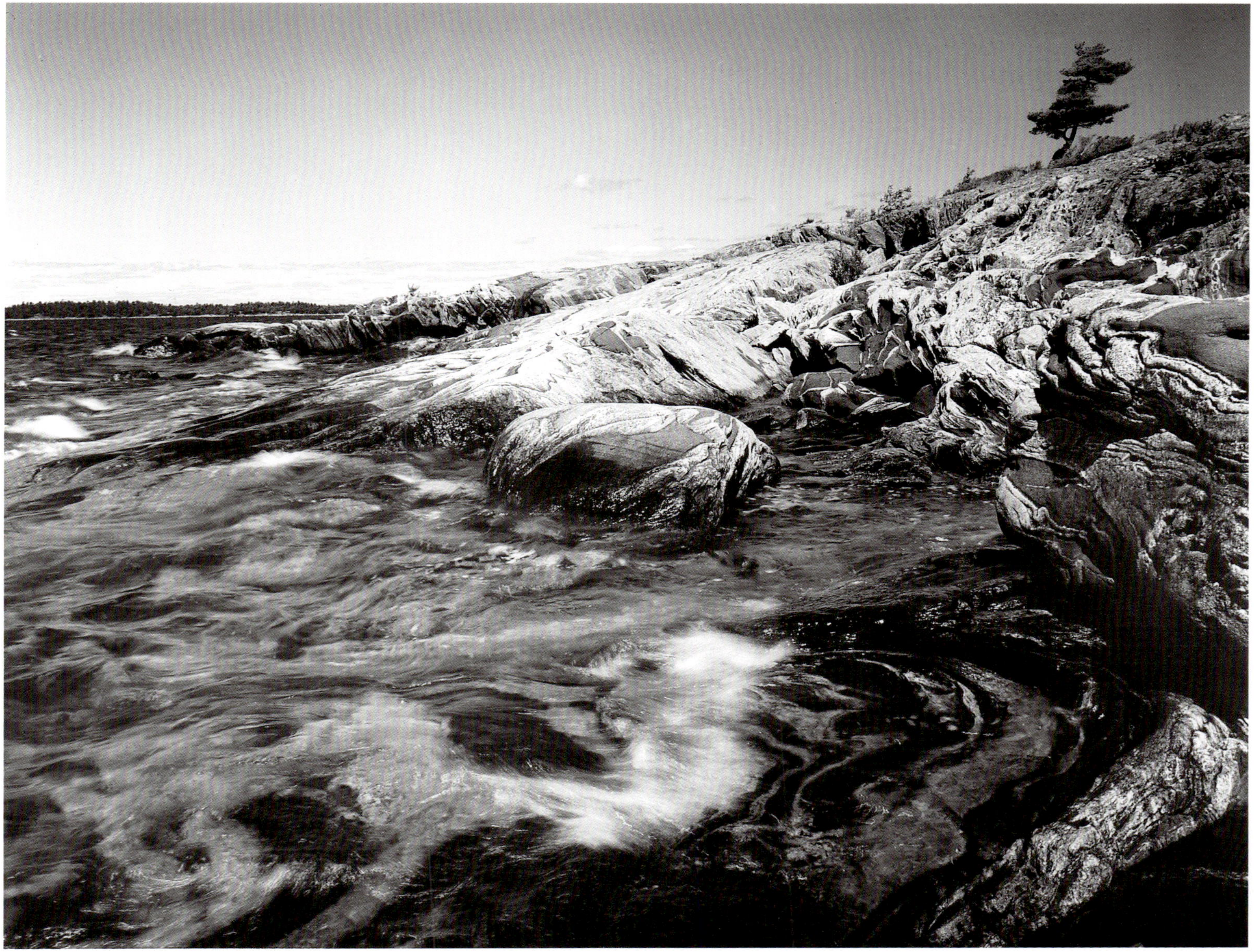

PLATE 5

DRIFTWOOD AND STONE, WRECK ISLAND   1993

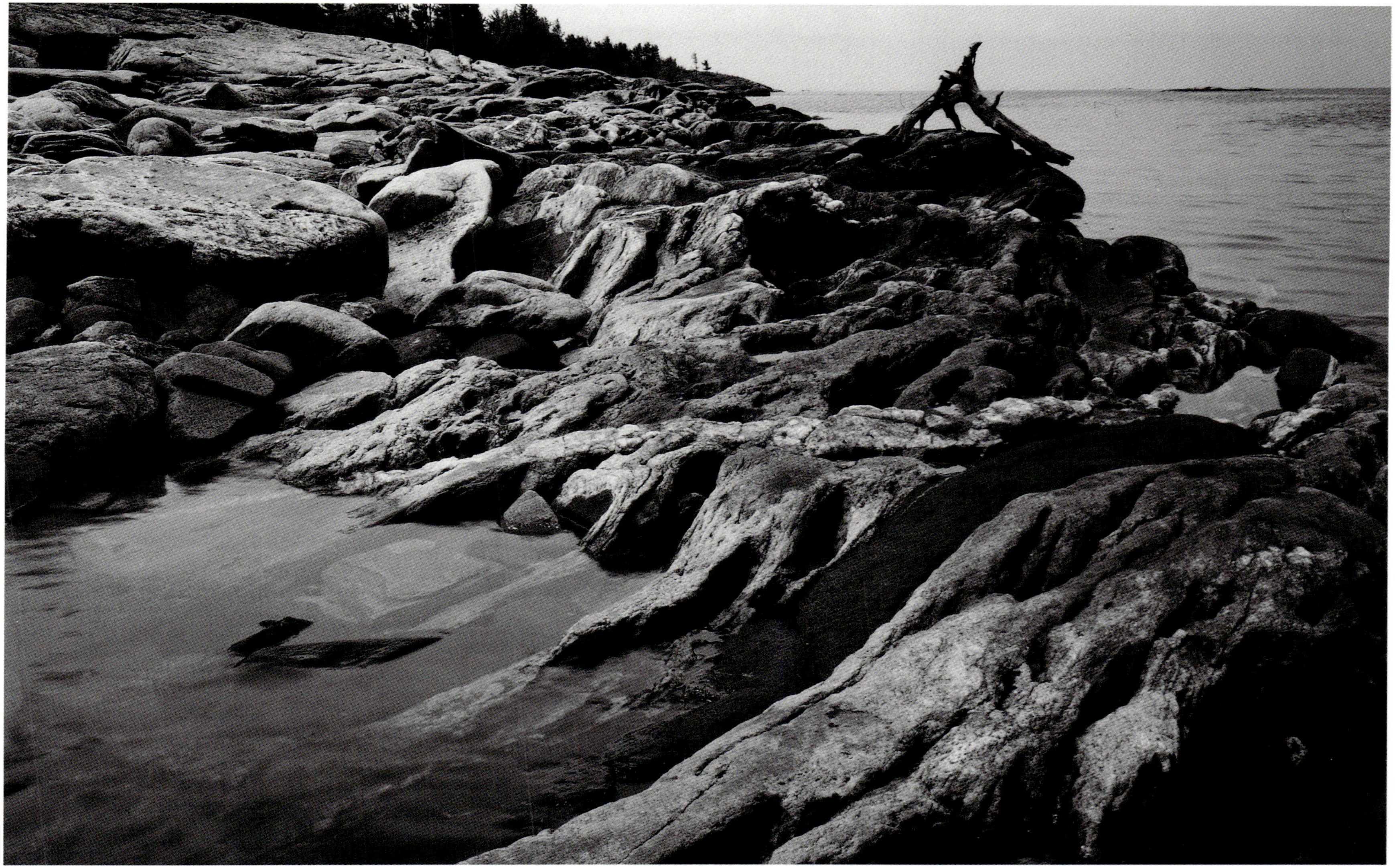

PLATE 6

STONES AND HORIZON, GEORGIAN BAY 1993

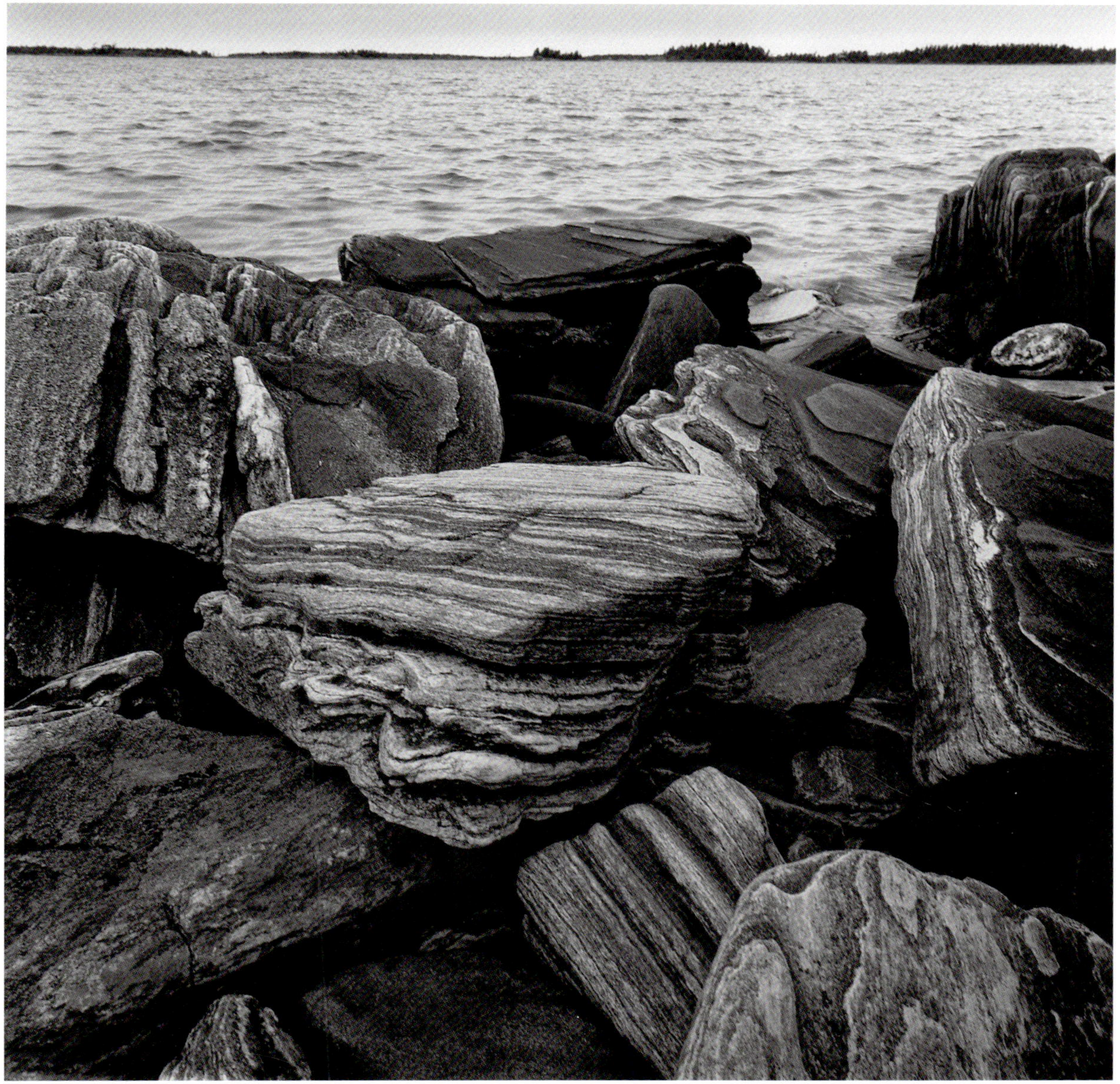

PLATE 7

FALLEN TREE, LA CLOCHE MOUNTAINS 1993

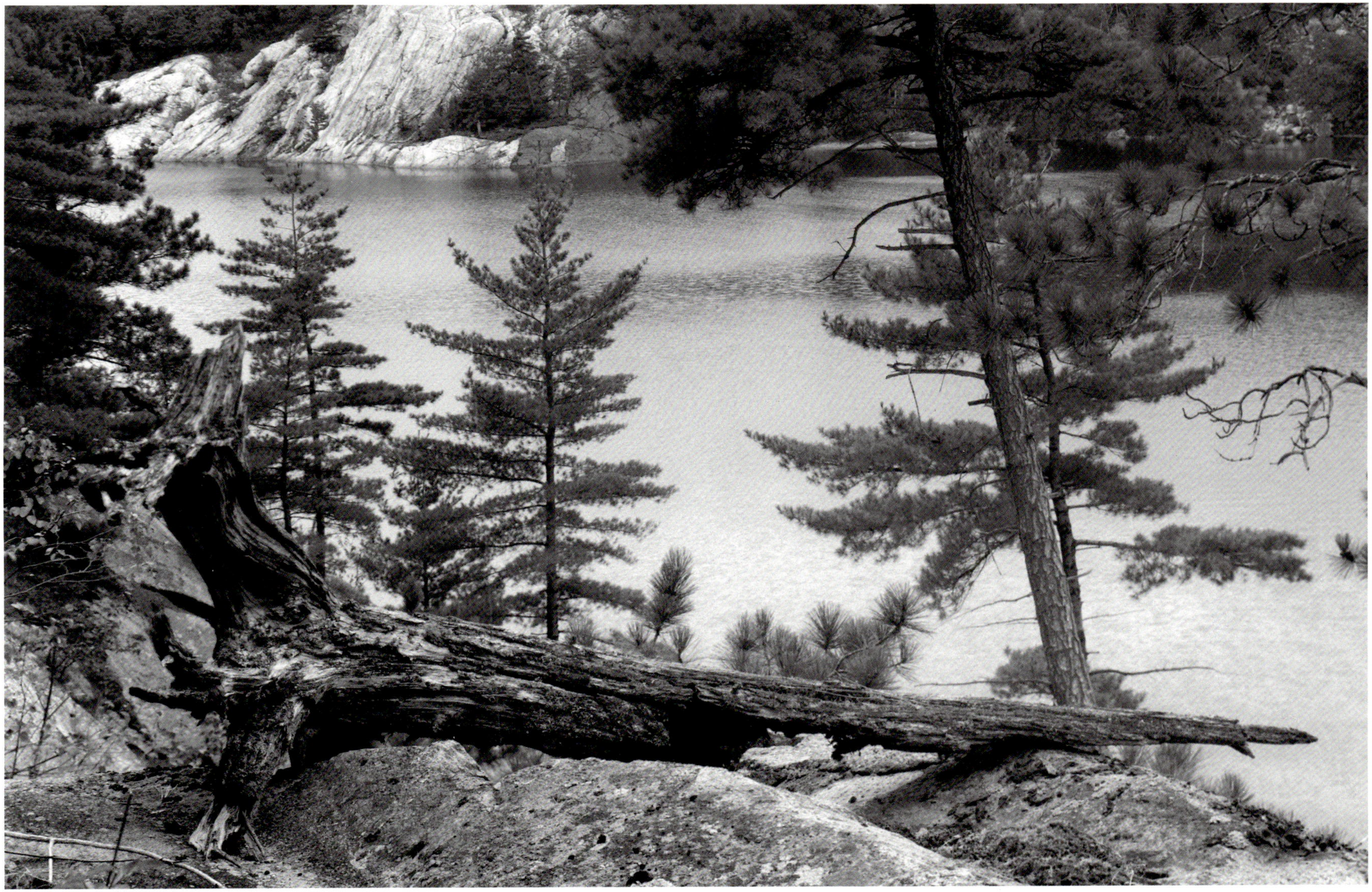

PLATE 8

MOULDED STONE, BUSTARD ISLANDS 1993

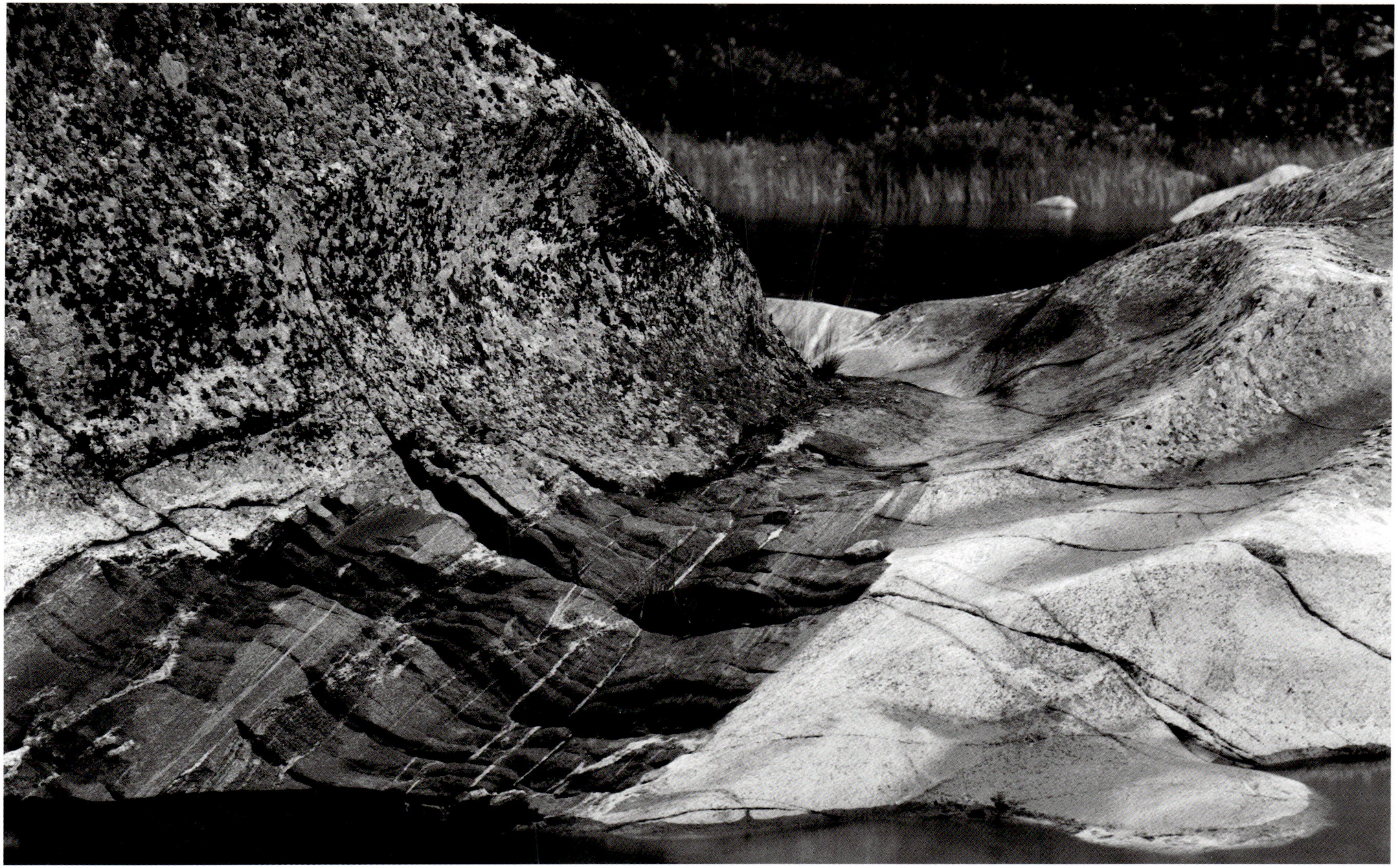

PLATE 9

FRENCH RIVER SHORELINE 1993

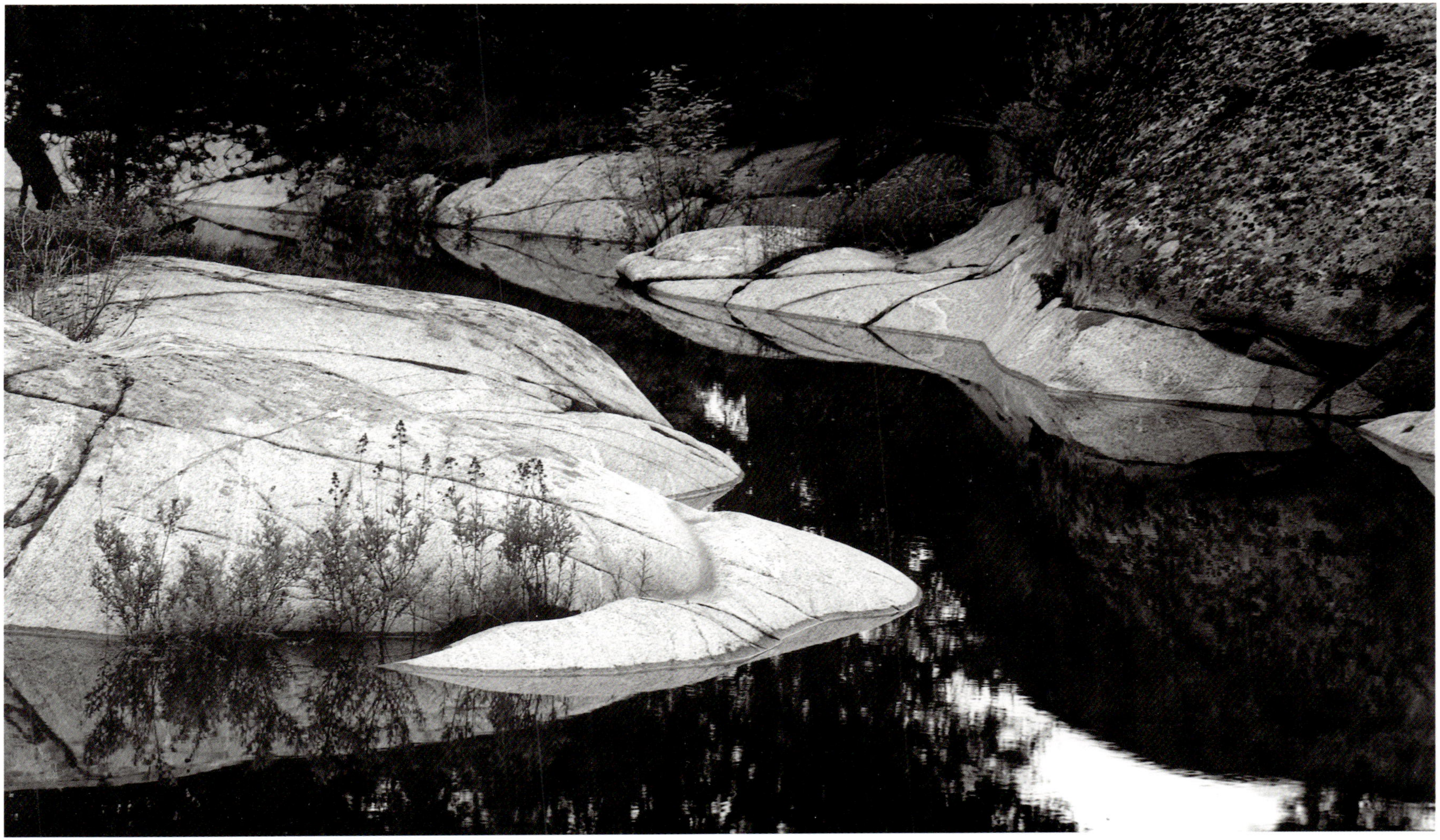

PLATE 10

FRENCH RIVER 1993

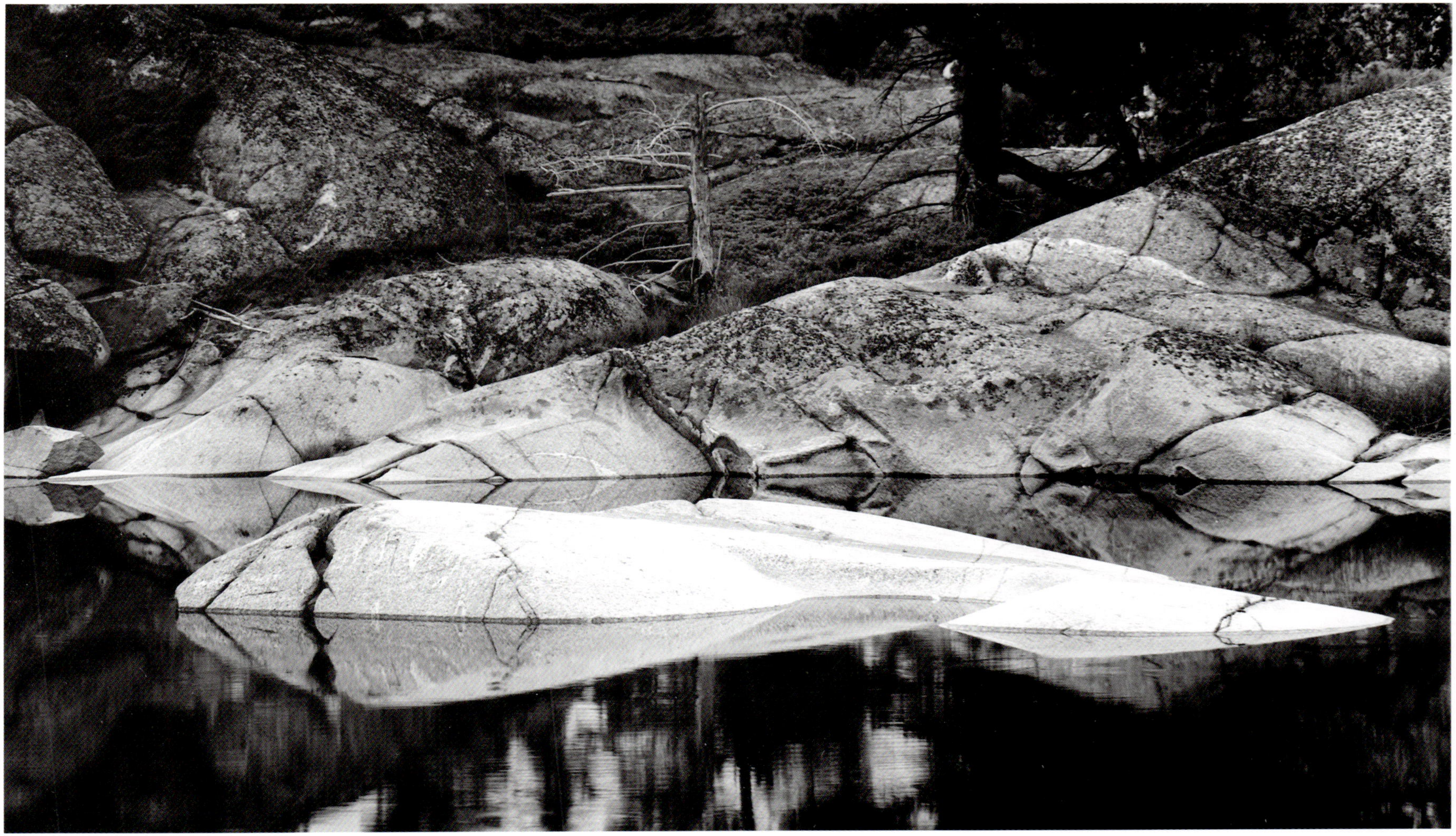

PLATE 11

VERTEBRAE, FRENCH RIVER 1993

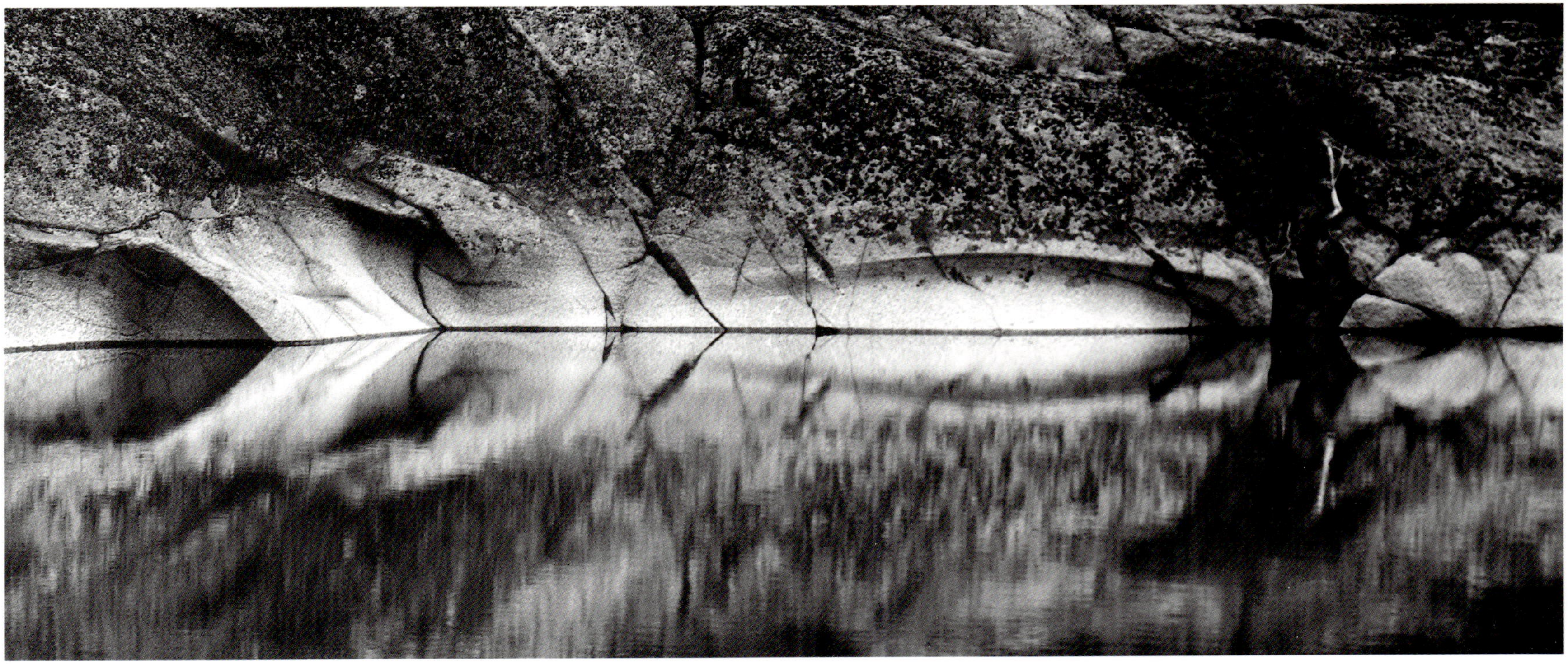

PLATE 12

DRAGONFLY, FRENCH RIVER 1993

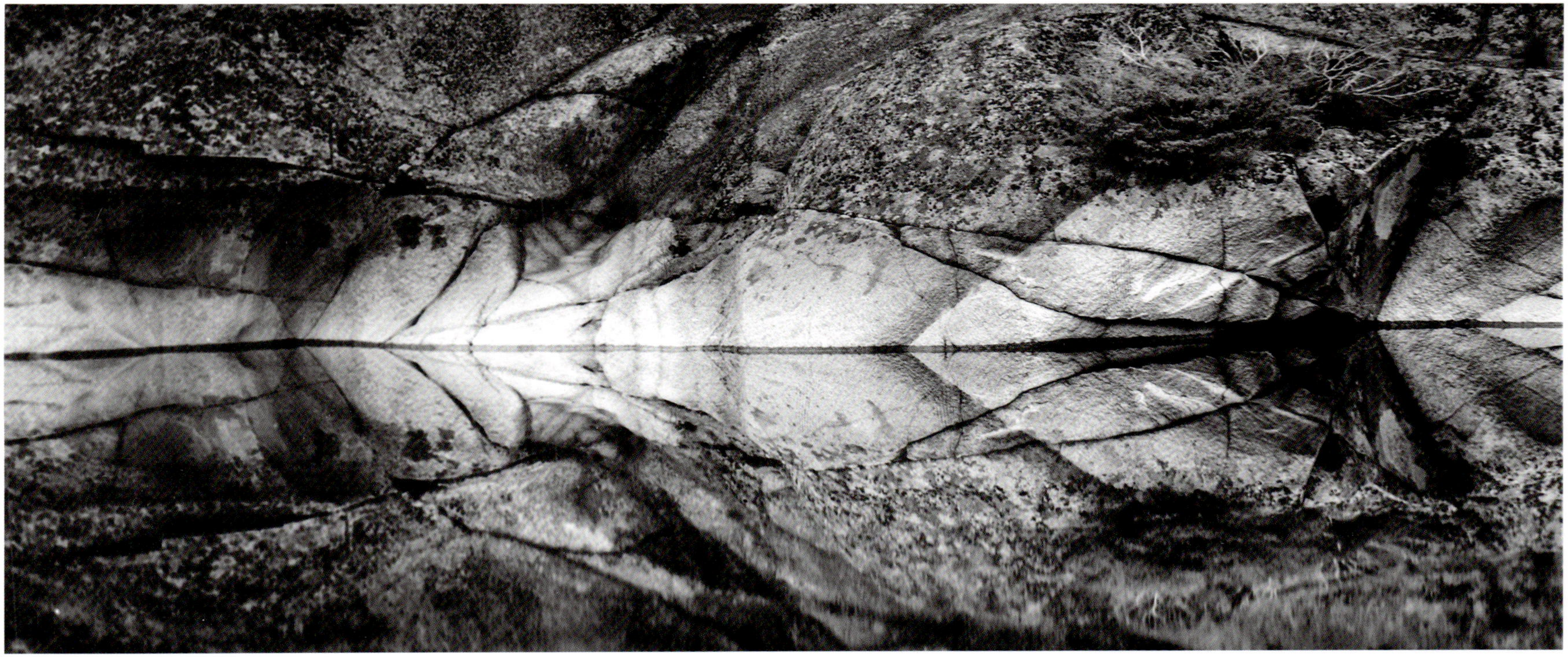

PLATE 13

ROCK AND PINE, GEORGIAN BAY 1993

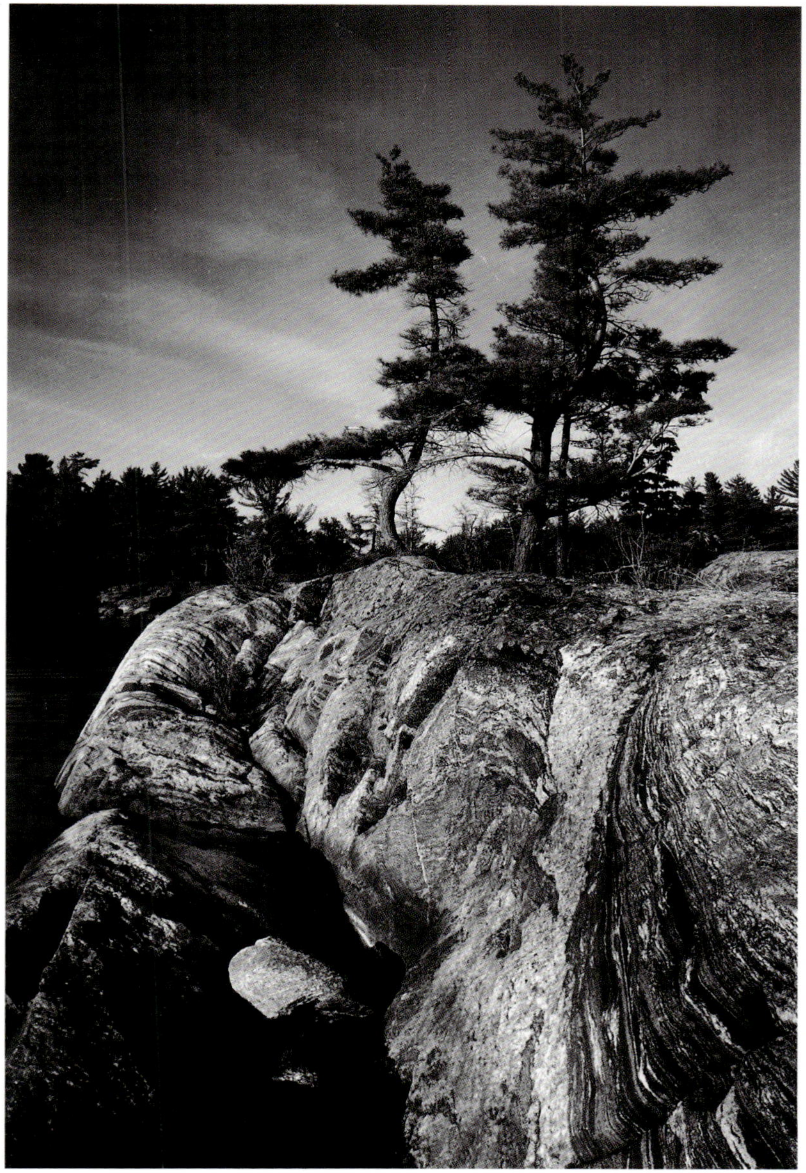

PLATE 14

FRENCH RIVER 1993

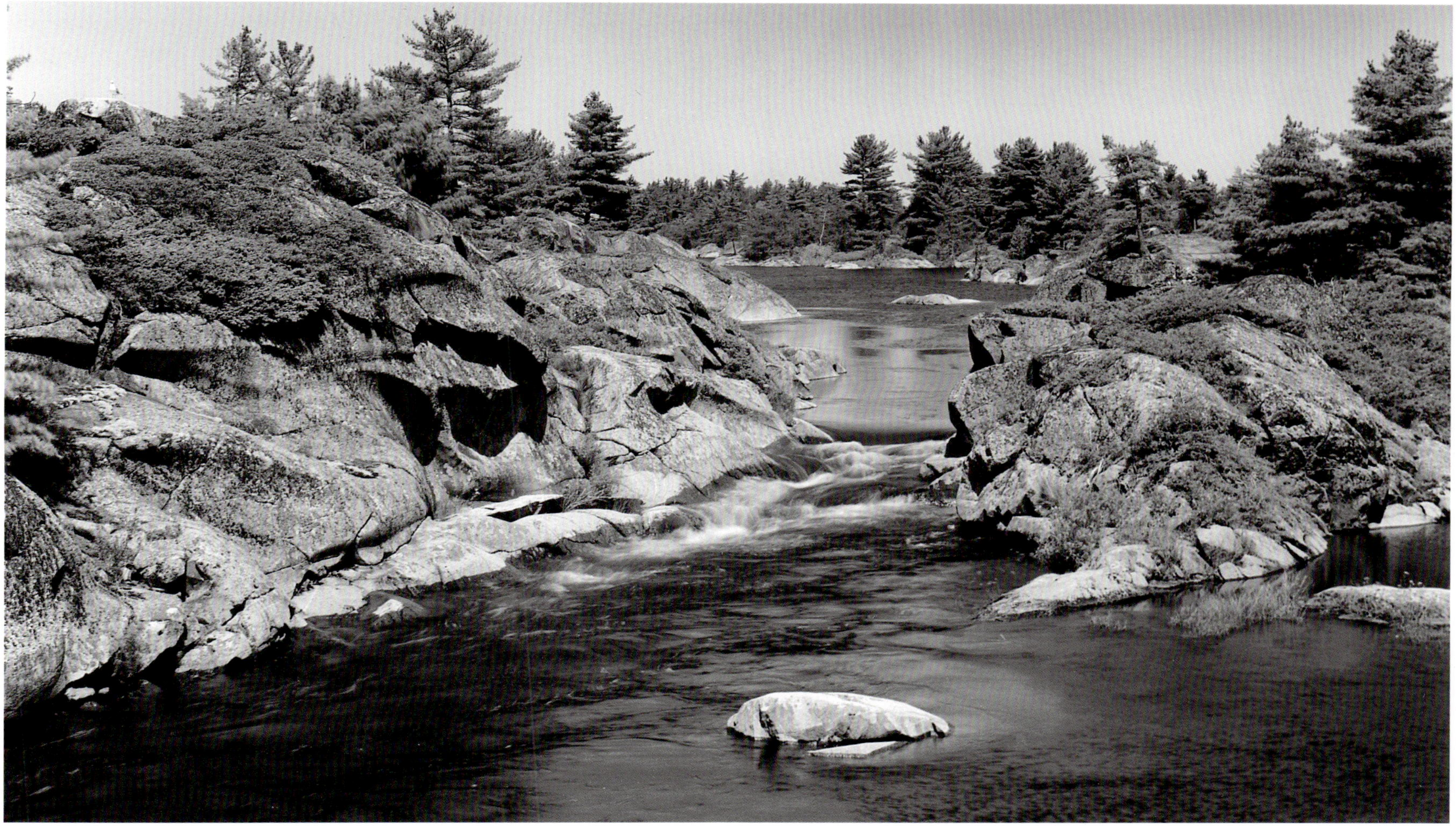

PLATE 15

PINE ROOTS, GEORGIAN BAY 1993

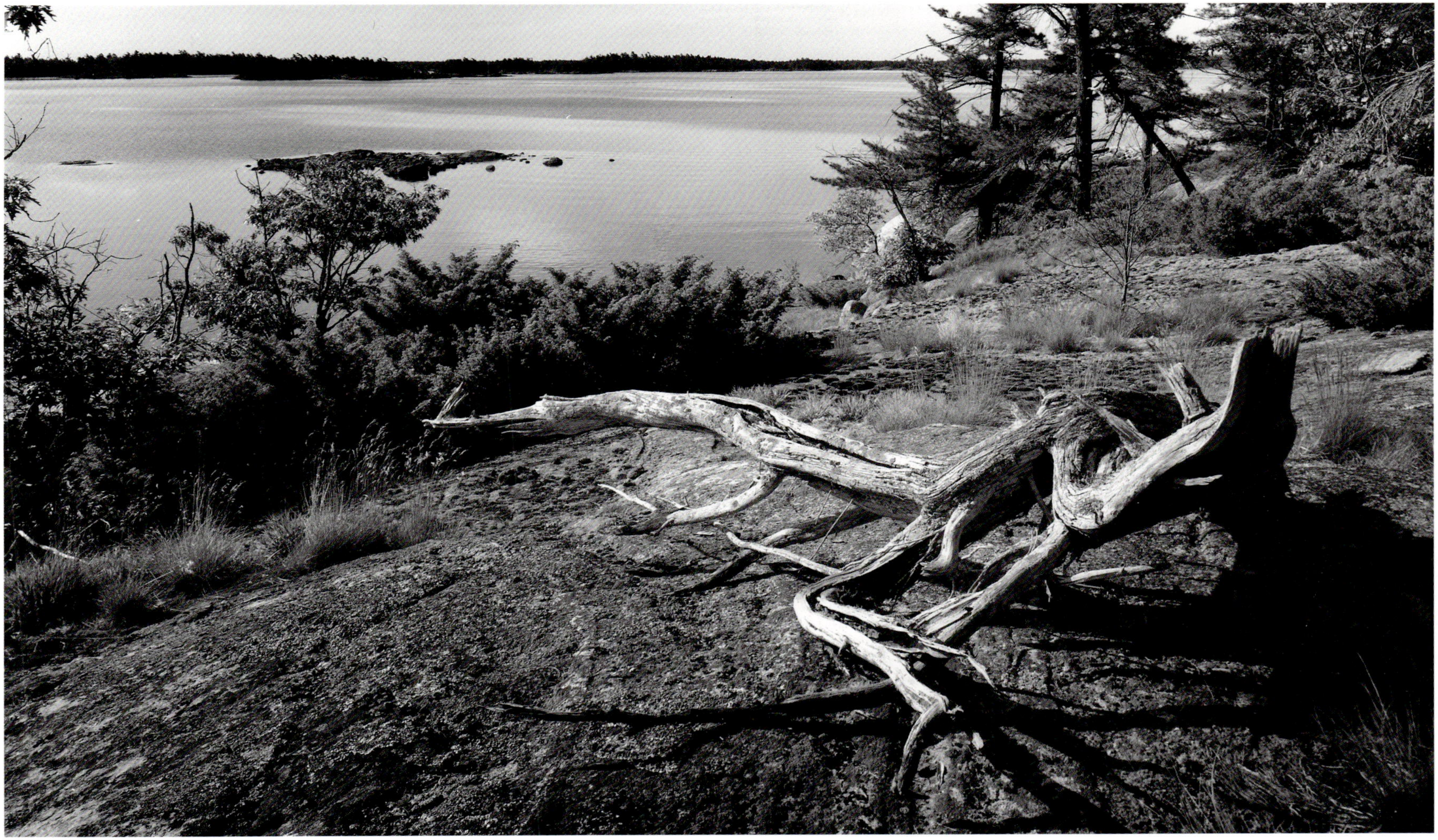

PLATE 16

BLOWING GRASS, EARLY MORNING, BUSTARD ISLANDS 1993

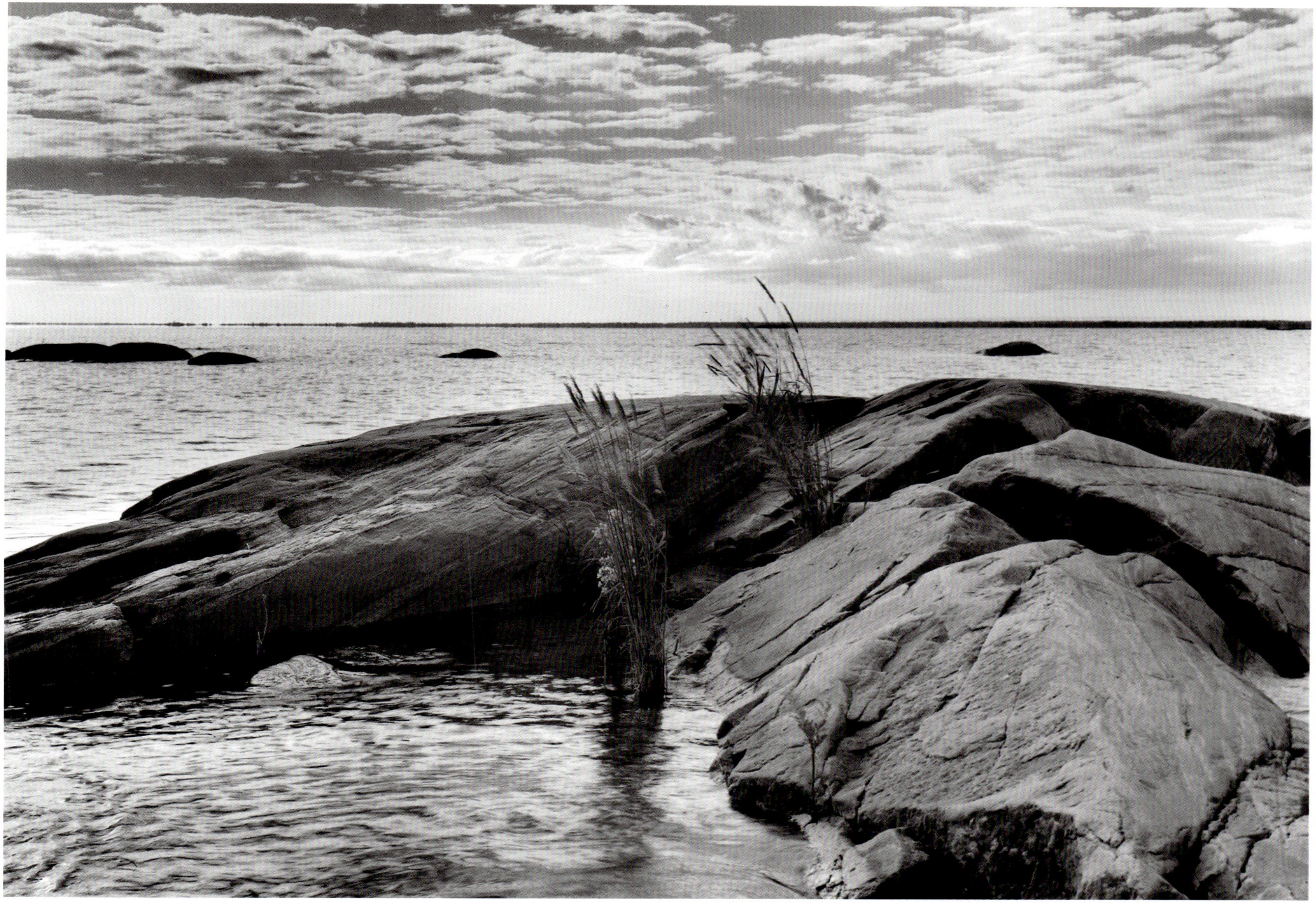

PLATE 17

TWISTED ROOTS, McGREGOR BAY 1992

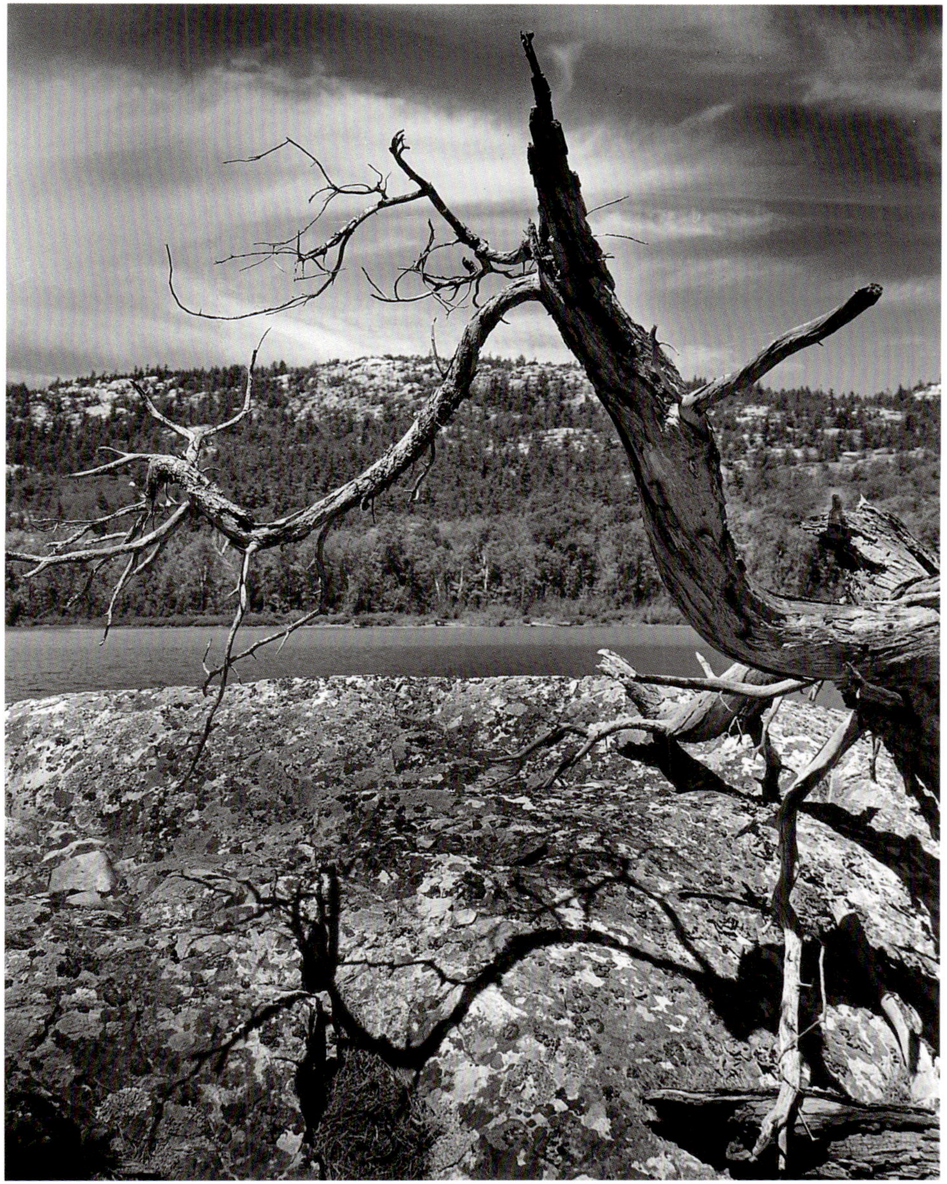

PLATE 18

BUSTARD ISLANDS SHORELINE 1993

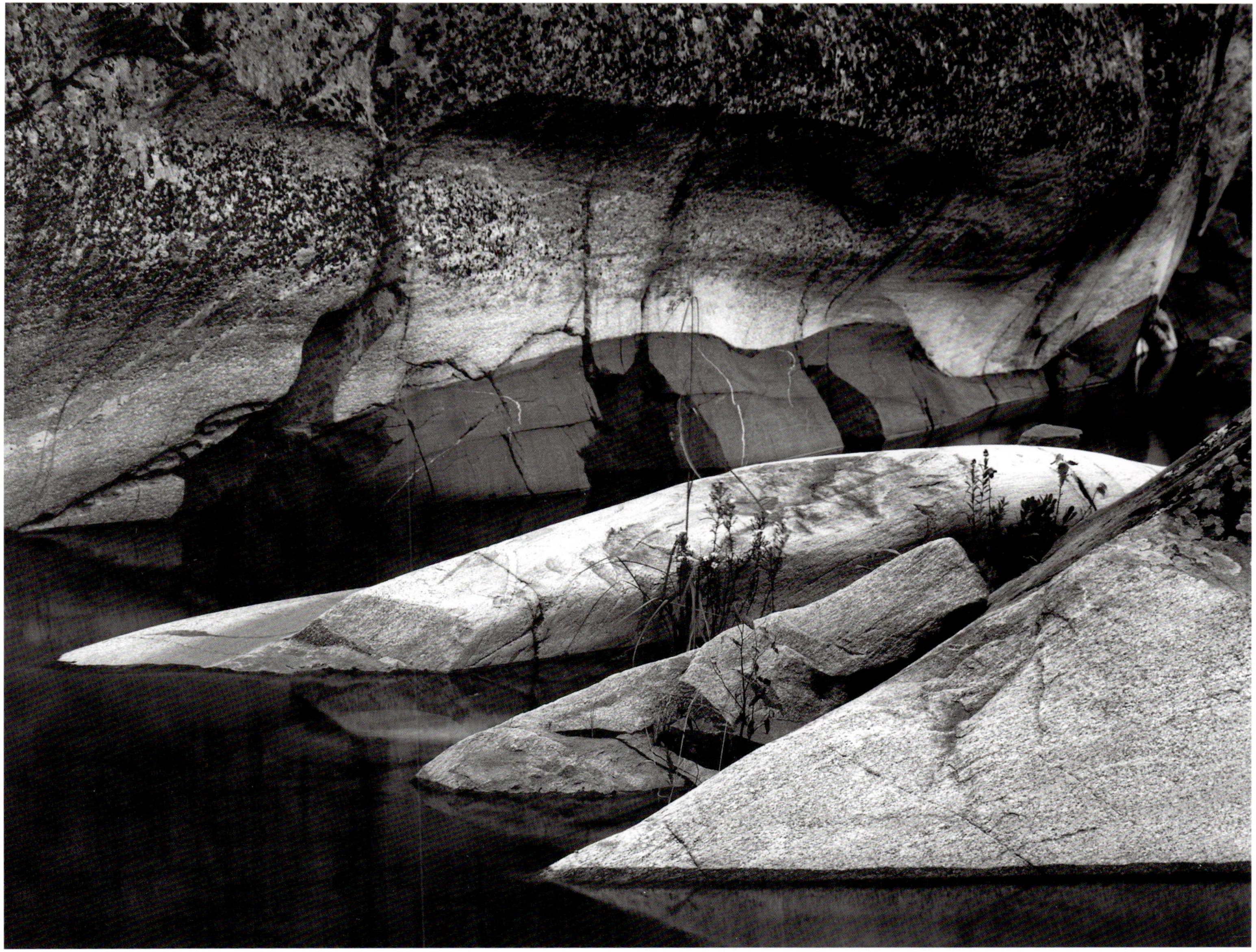

PLATE 19

STREAM NEAR HIGH FALLS, MUSKOKA RIVER 1992

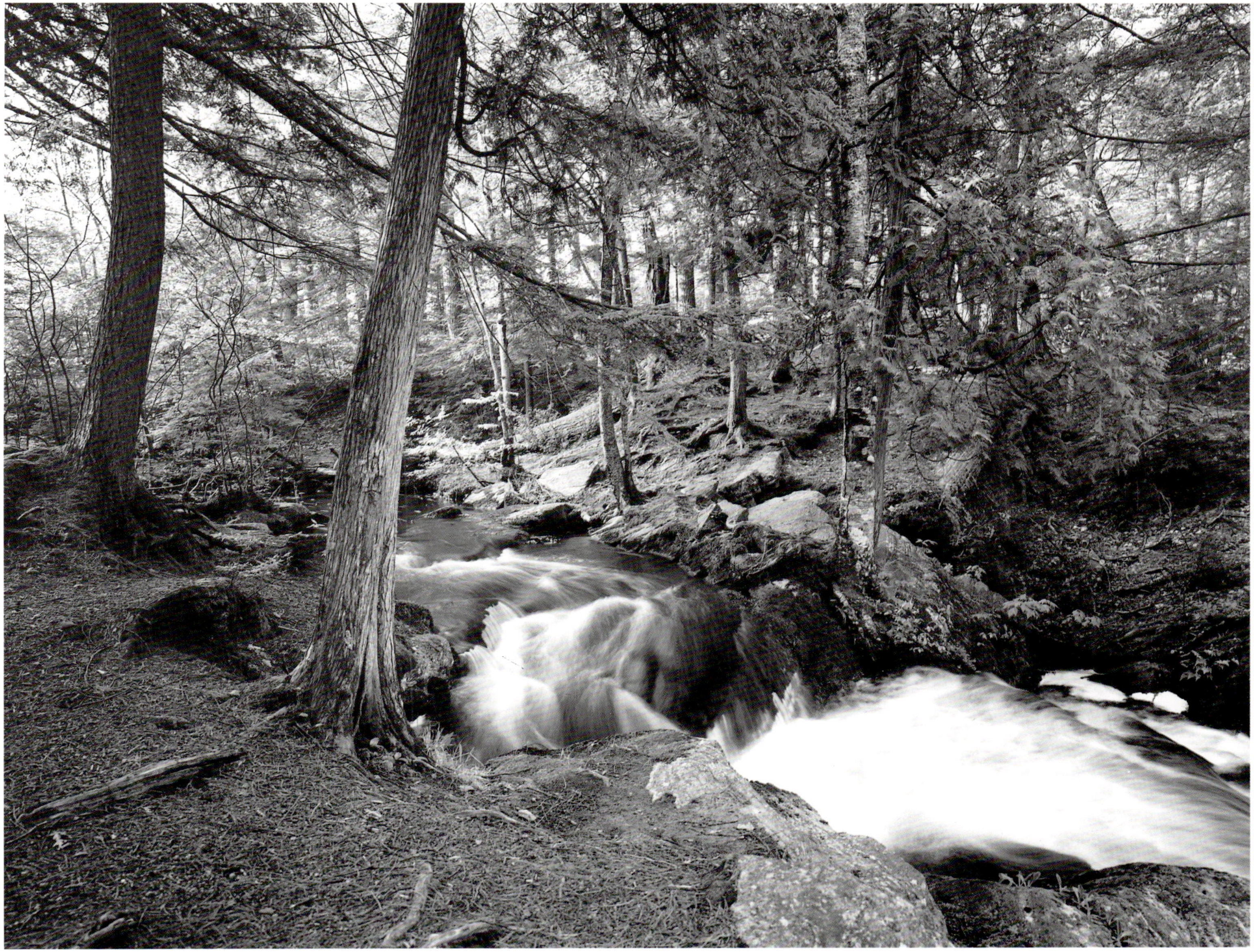

PLATE 20

BASS ROCK, BRACEBRIDGE 1992

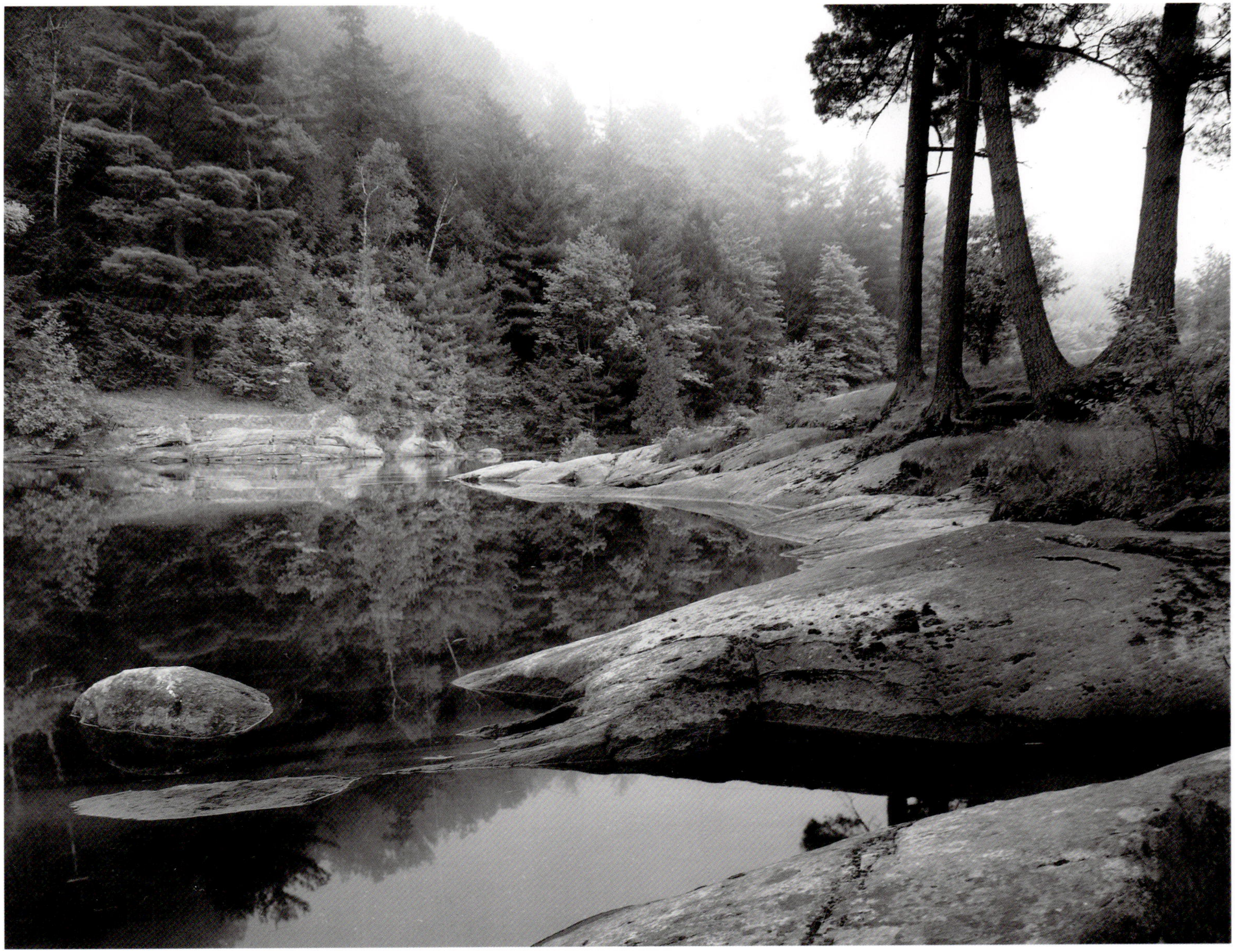

PLATE 21

ROCK LEDGES, MUSKOKA RIVER 1992

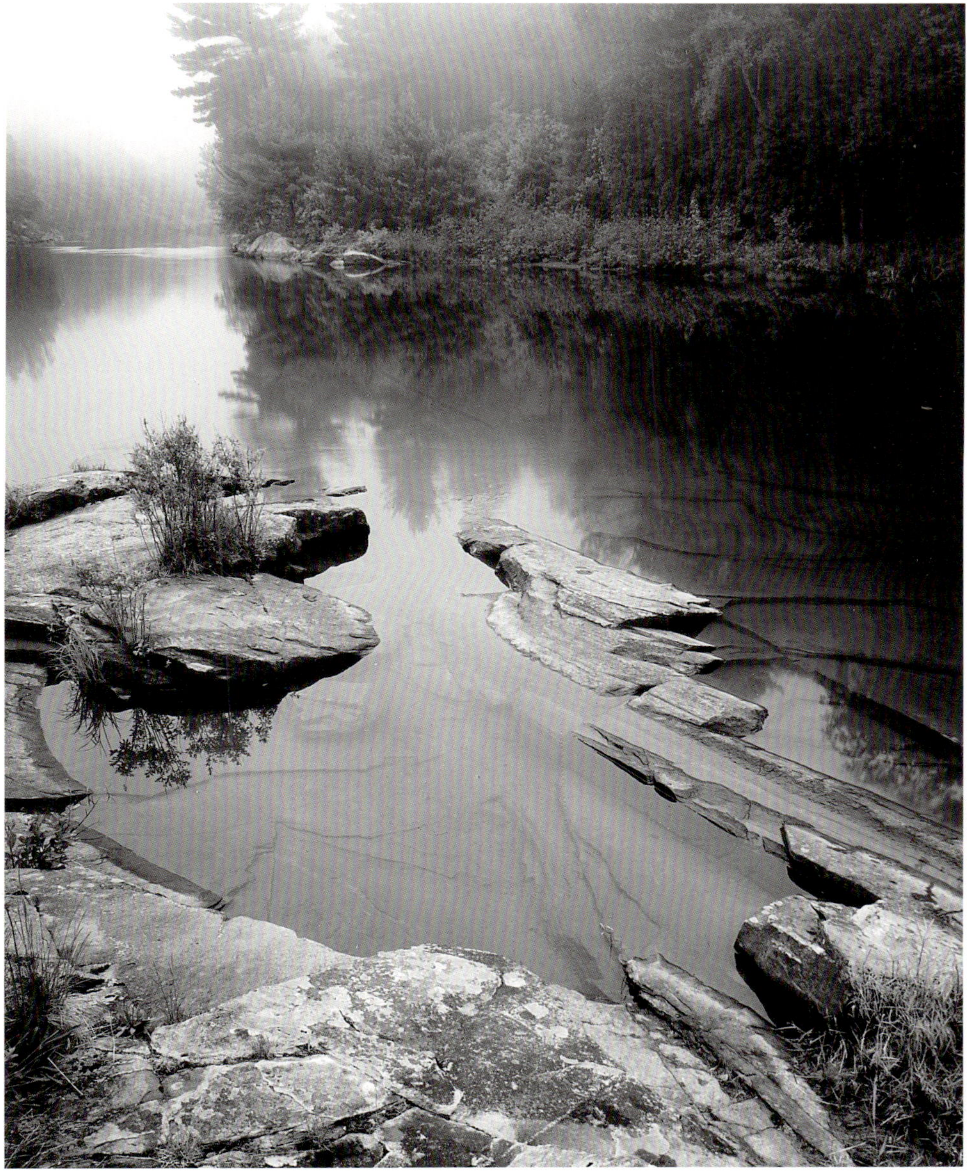

PLATE 22

RAGGED FALLS, OXTONGUE RIVER 1993

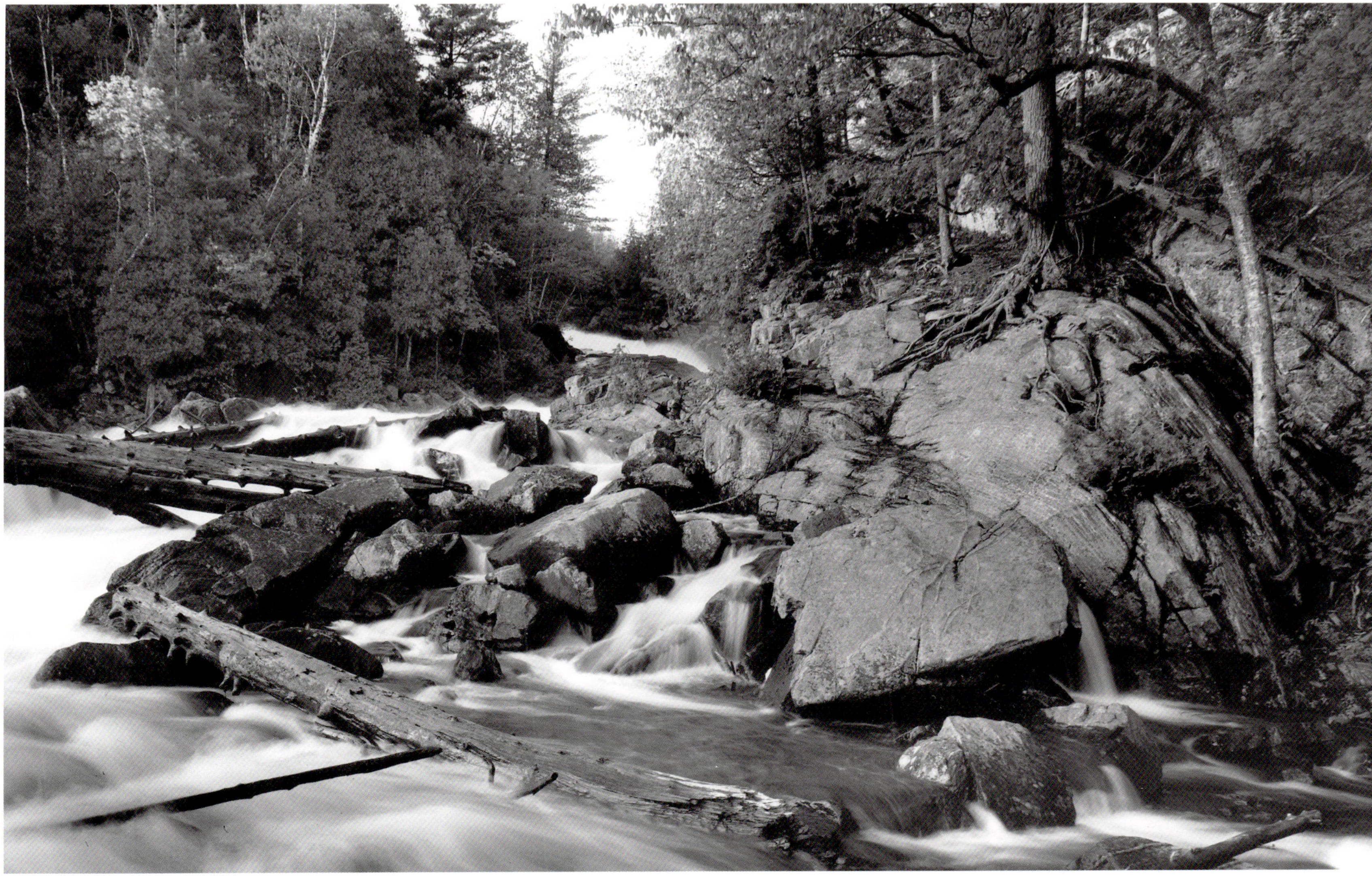

PLATE 23

STONE AT RAGGED FALLS 1993

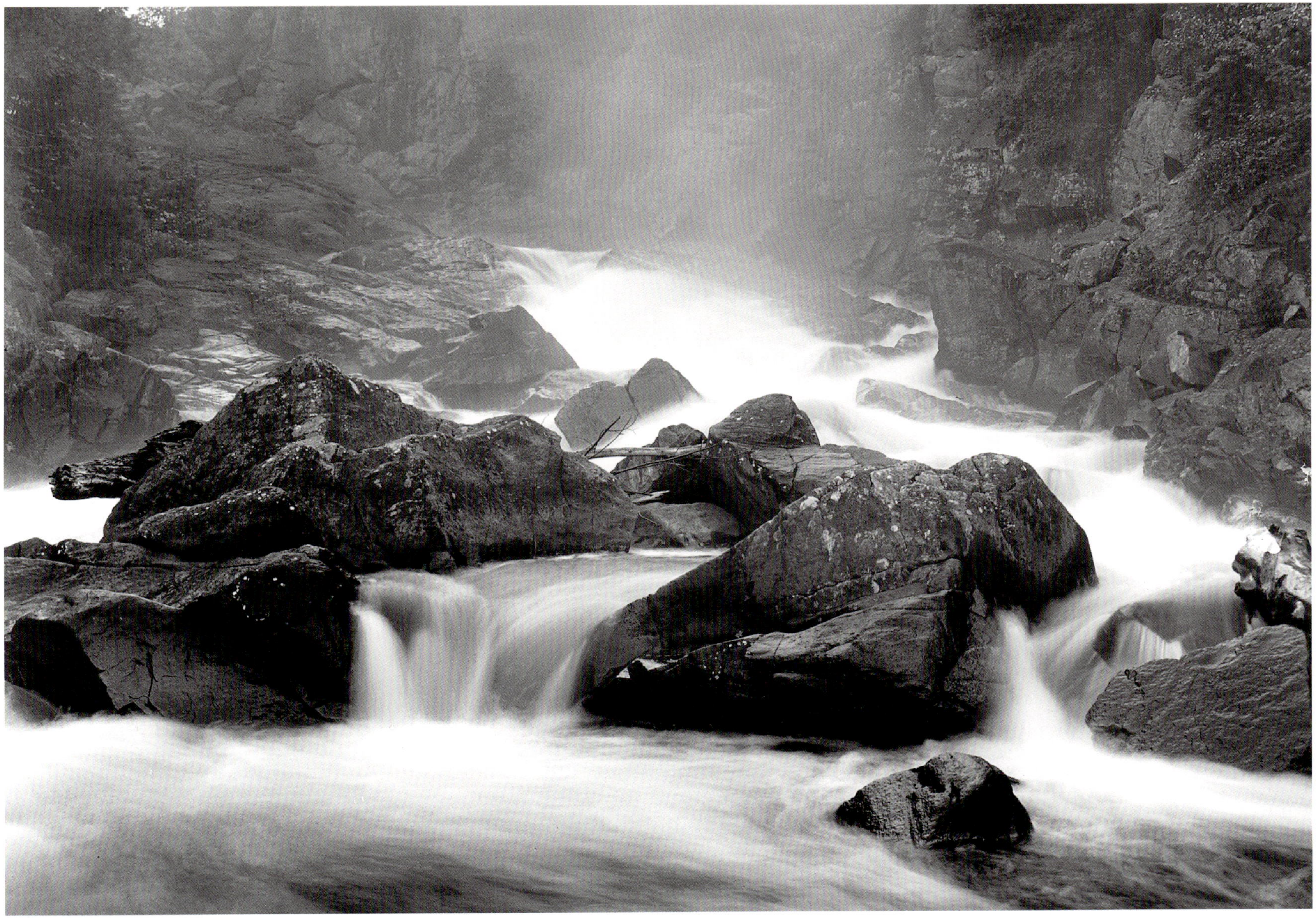

PLATE 24

OXTONGUE RAPIDS 1993

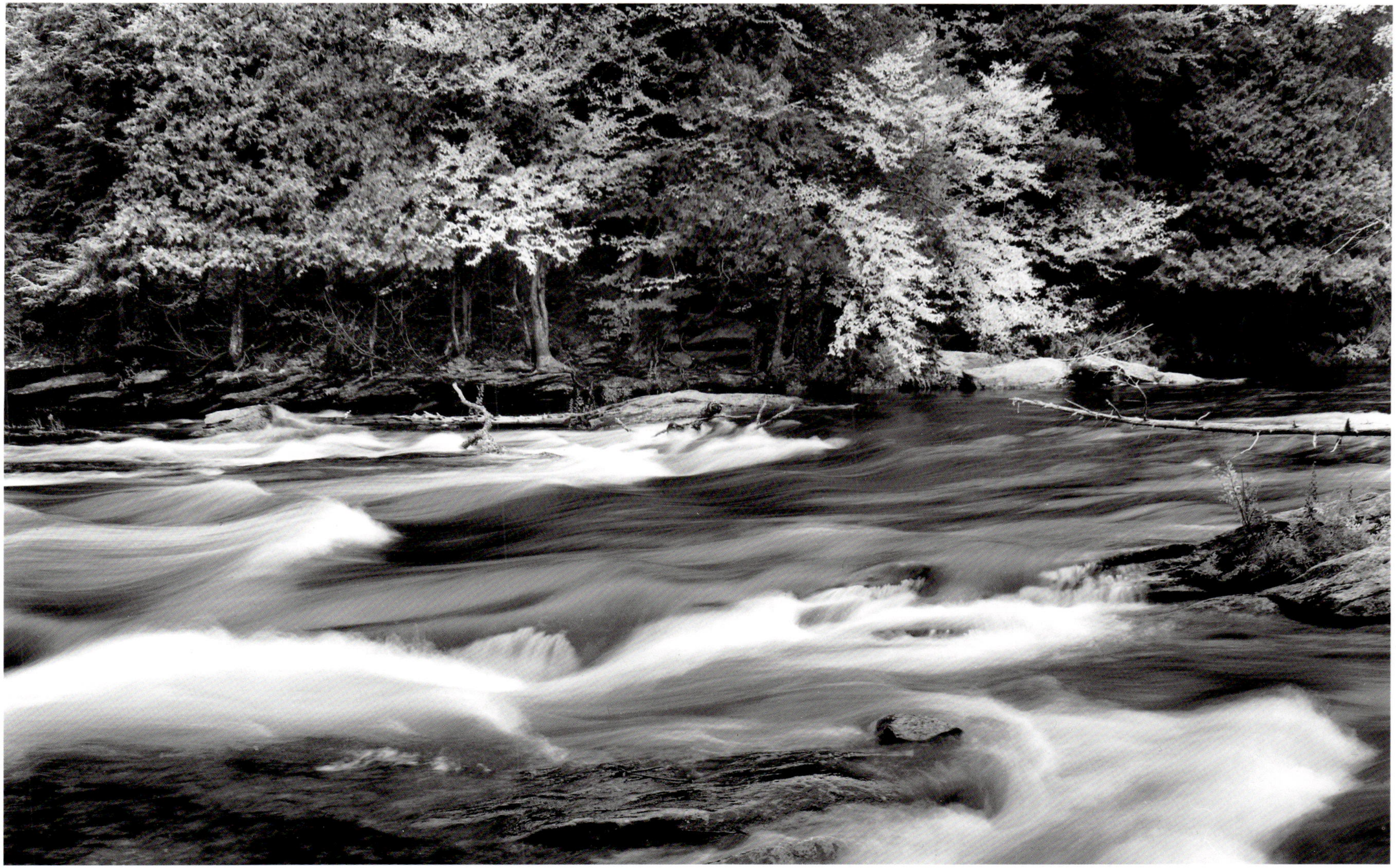

PLATE 25

HUCKLEBERRY RIDGE, MILFORD BAY 1992

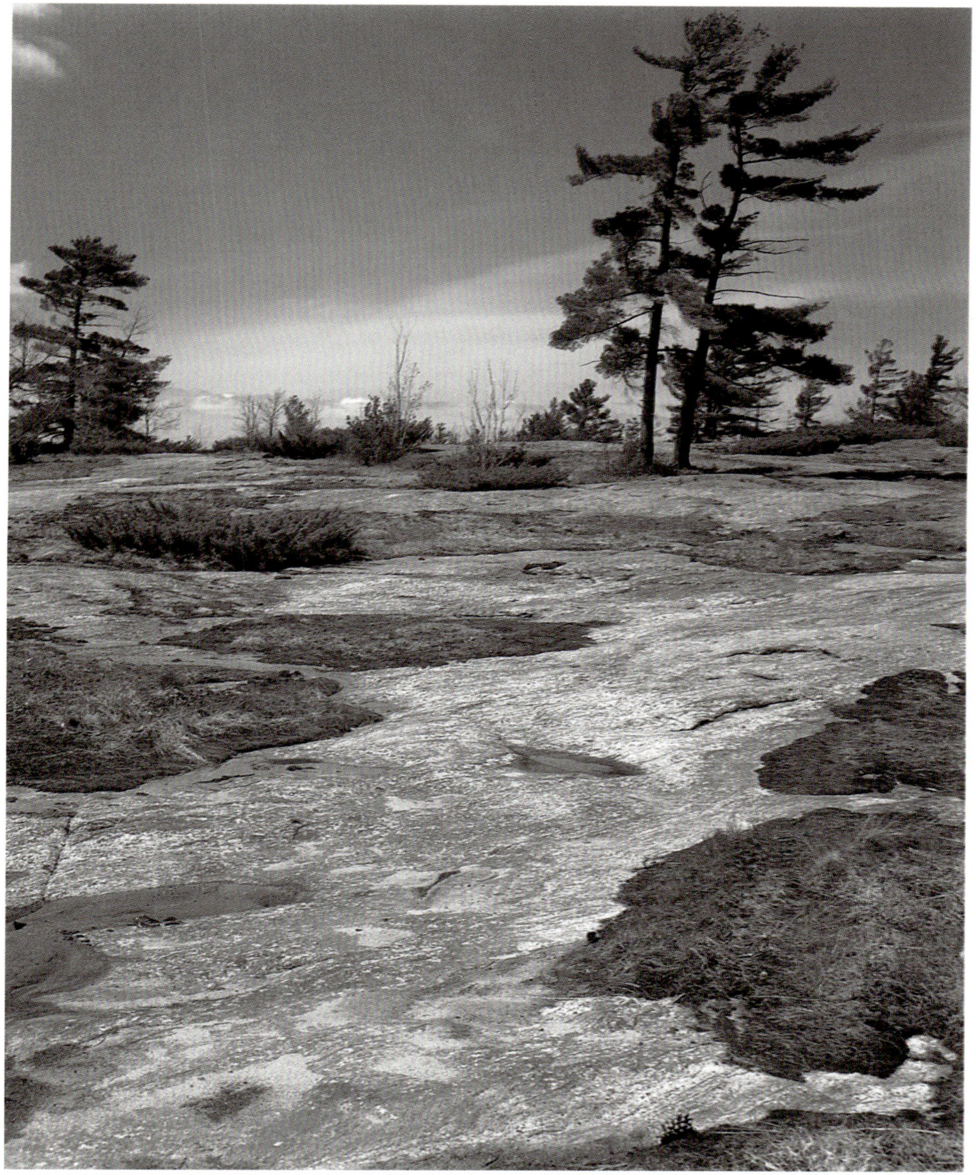

PLATE 26

ROOTS AND TEXTURE NEAR MUSKOKA RIVER, BRACEBRIDGE 1993

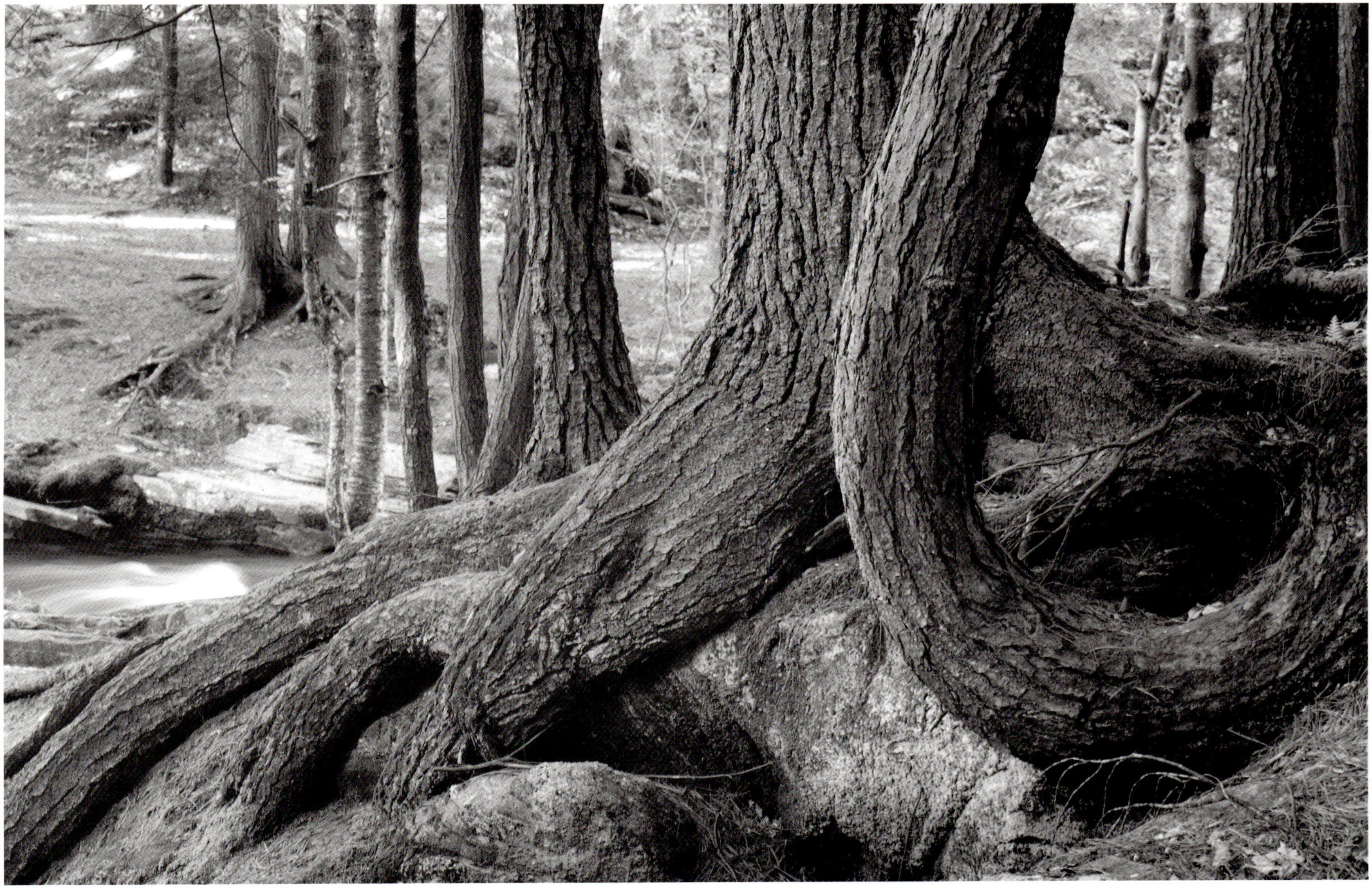

PLATE 27

BEAVER POND NEAR LAKE MUSKOKA 1992

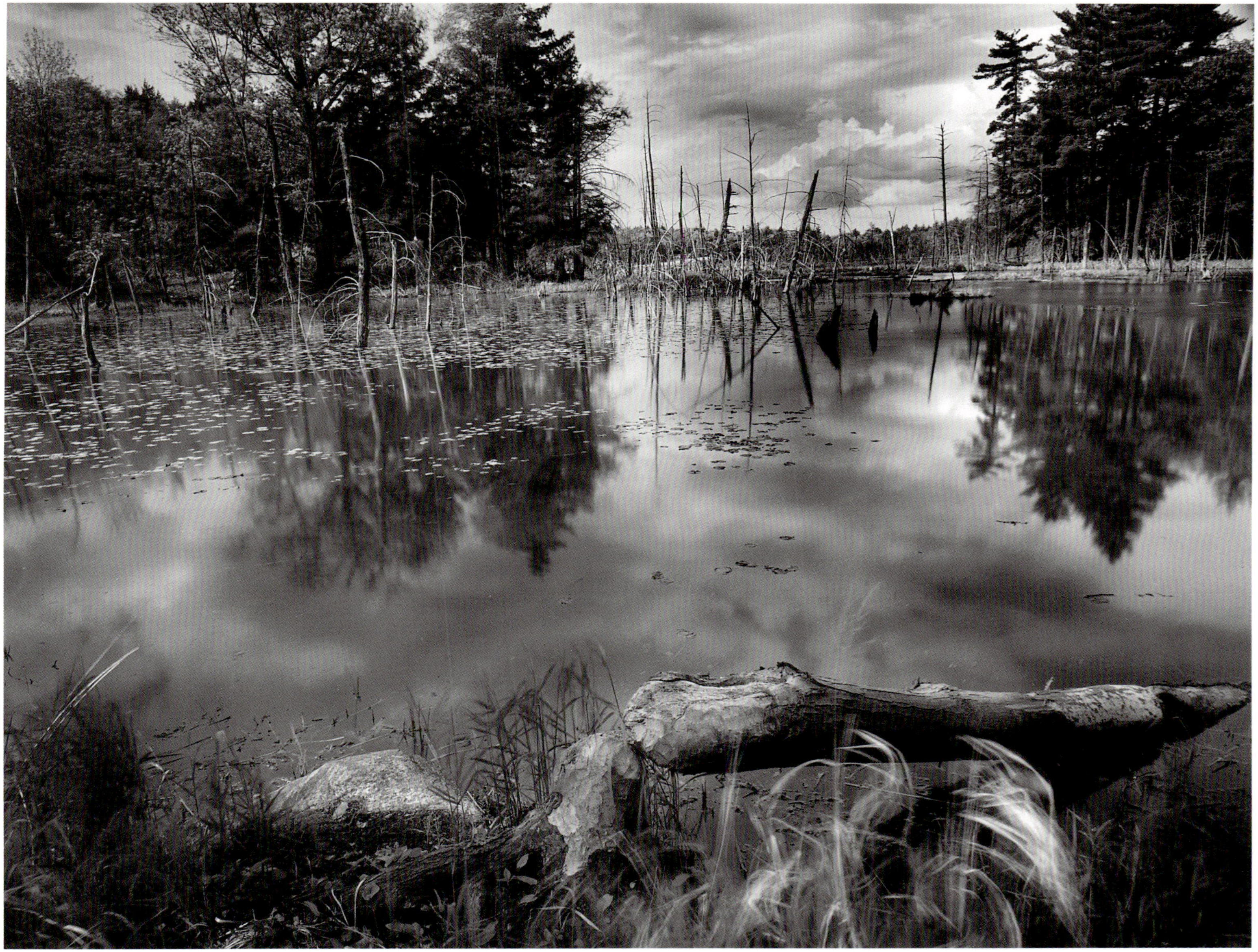

PLATE 28

SWAMP NEAR LAKE MUSKOKA 1992

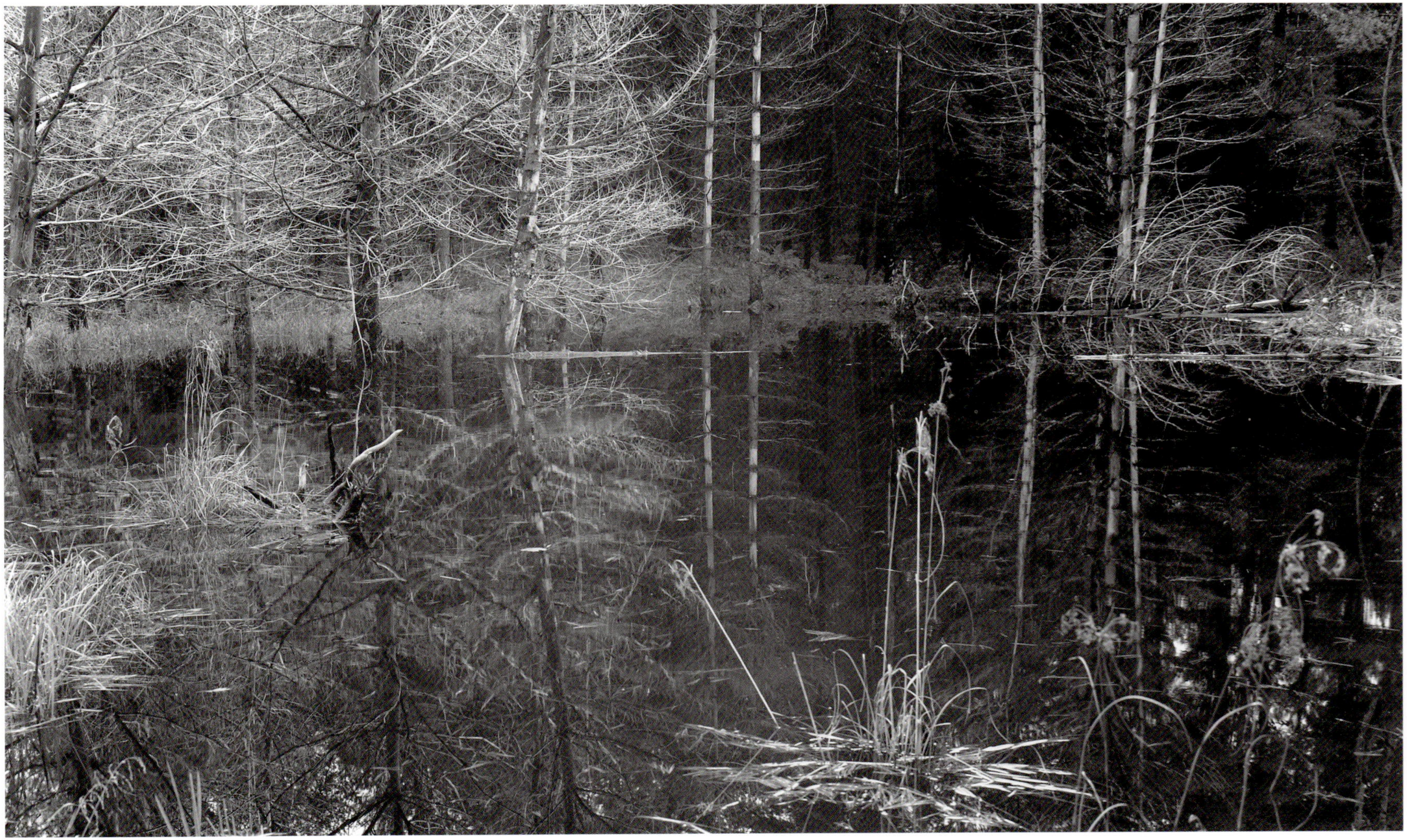

PLATE 29

SWAMP NEAR BRACEBRIDGE 1993

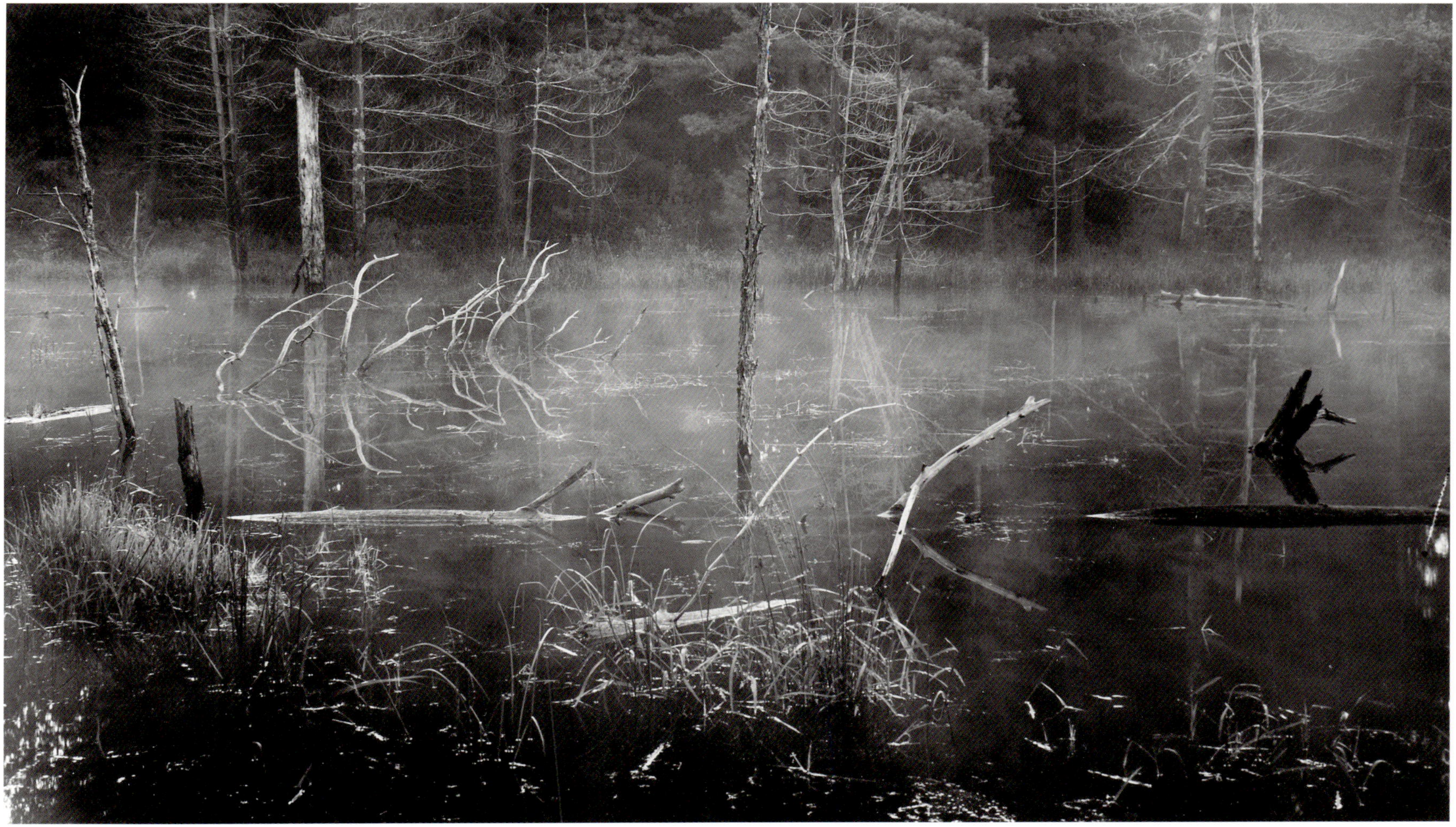

PLATE 30

CREEK AT HIGH FALLS, MUSKOKA RIVER 1993

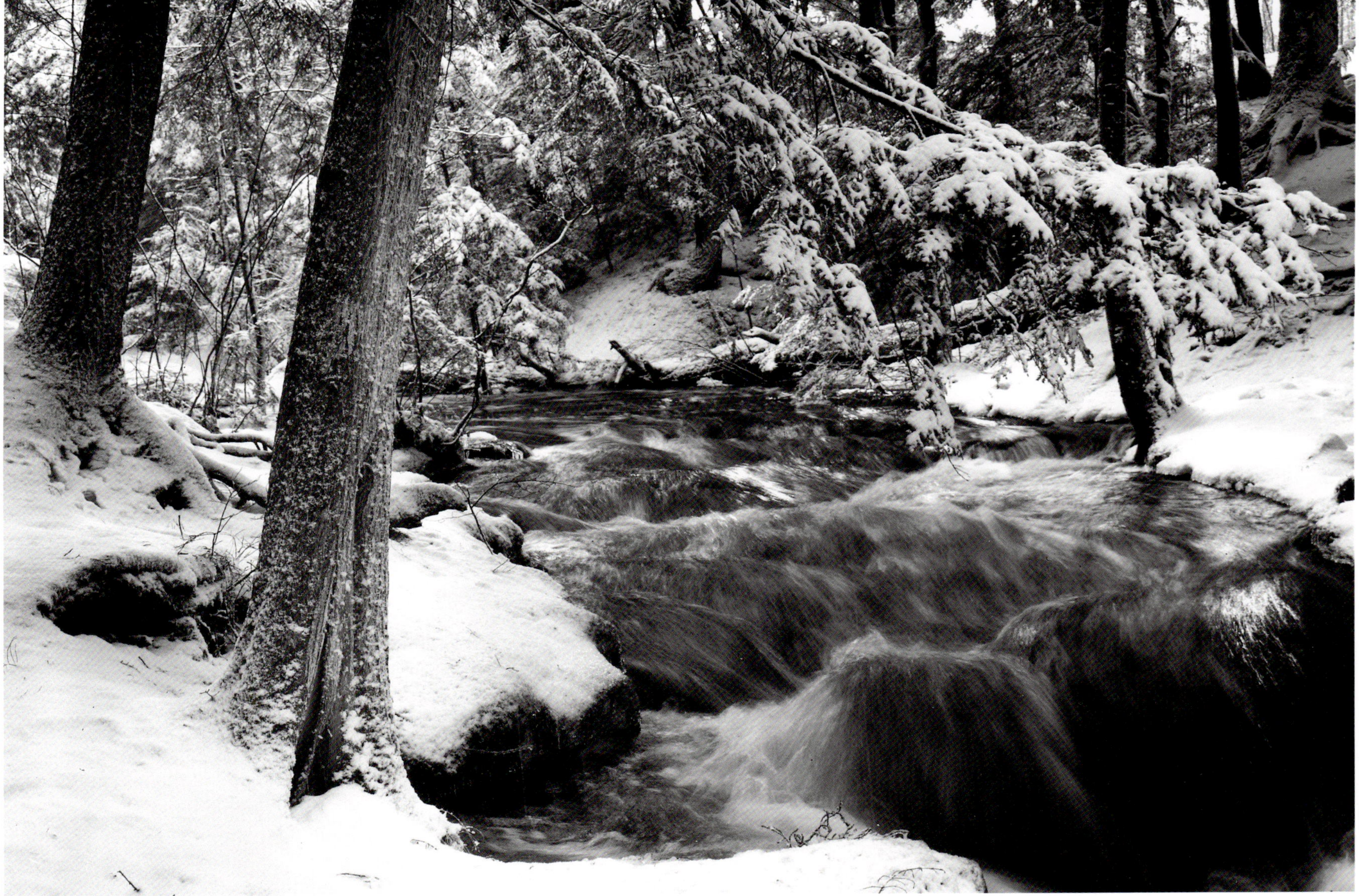

PLATE 31

FIRST SNOW, CAMP HOLLYBURN 1992

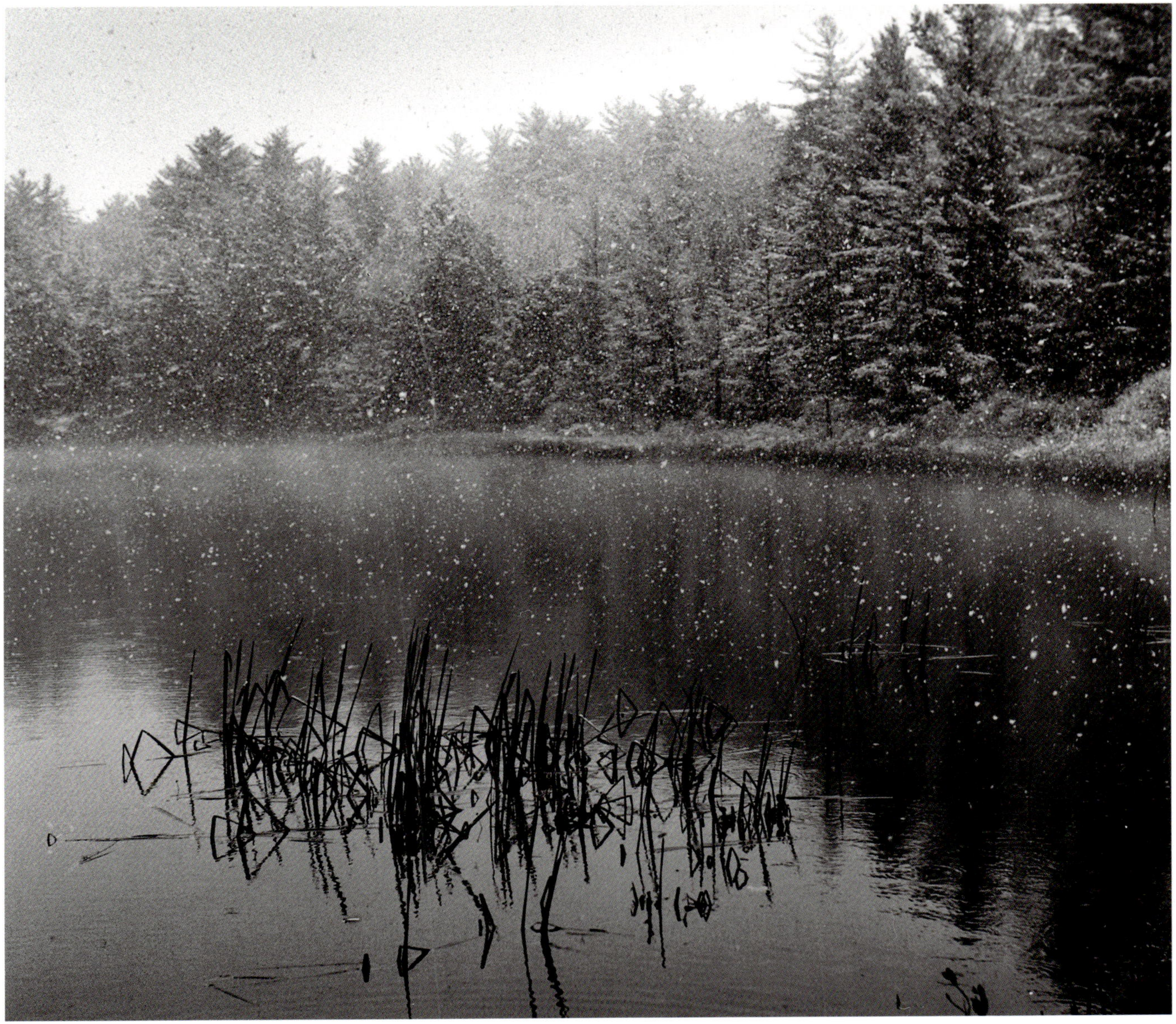

PLATE 32

SNOW AND CATTAILS, BAYSVILLE HIGHWAY 1994

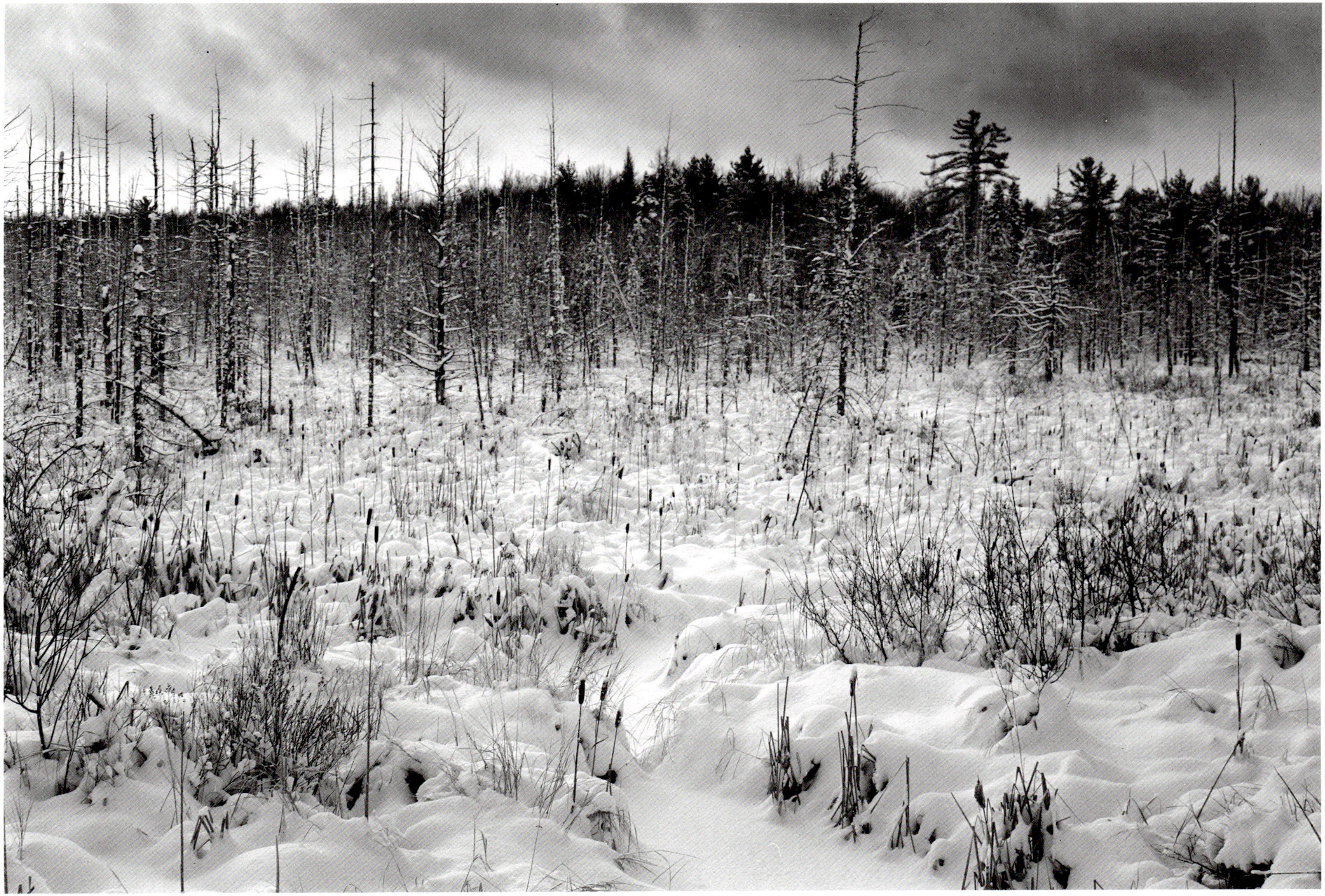

PLATE 33

SNOW AND WATER, BRACEBRIDGE 1992

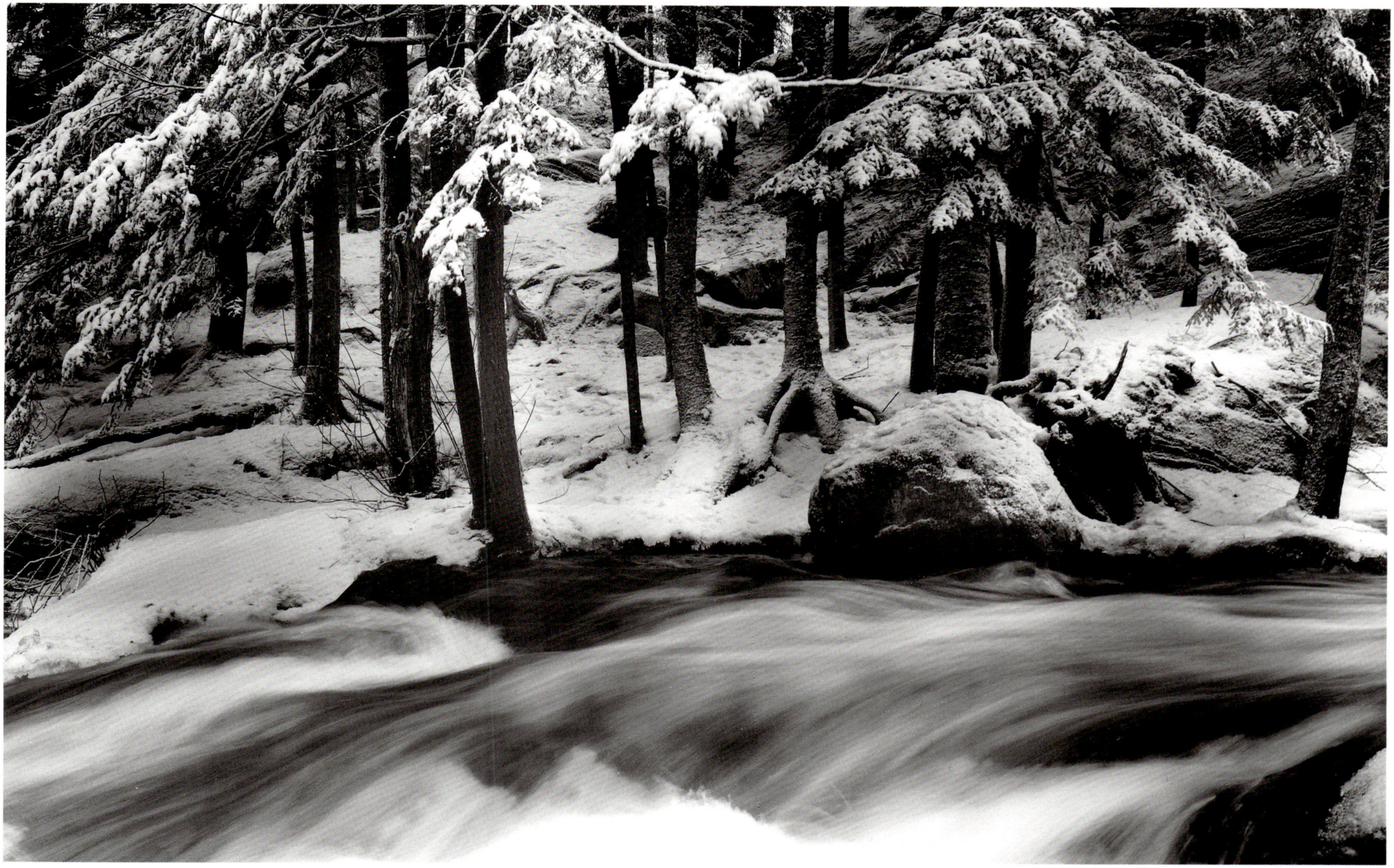

PLATE 34

SNOW AND LIGHT NEAR PORT SYDNEY 1993

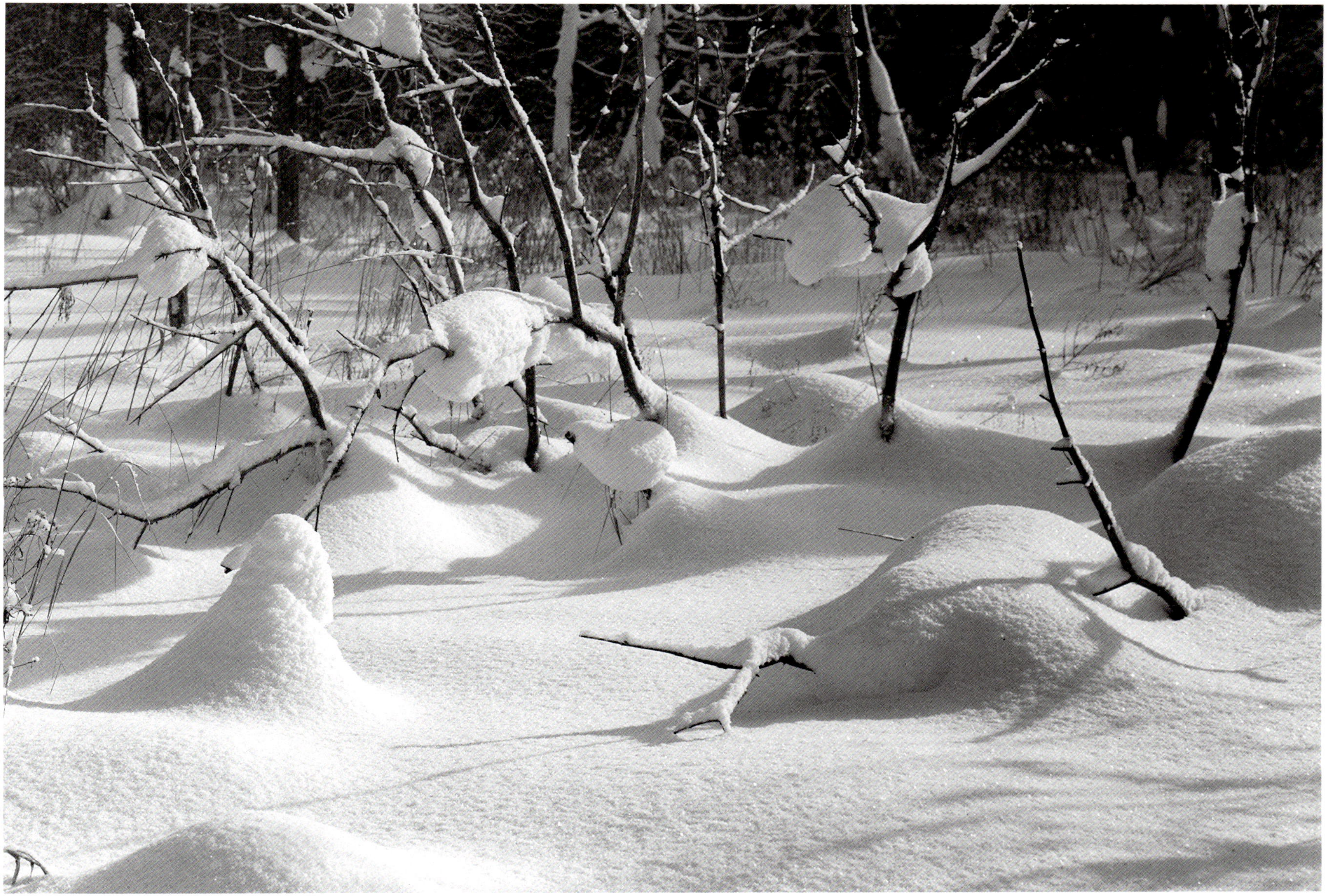

PLATE 35

SNOW AND GRASS, VANKOUGHNET ROAD 1994

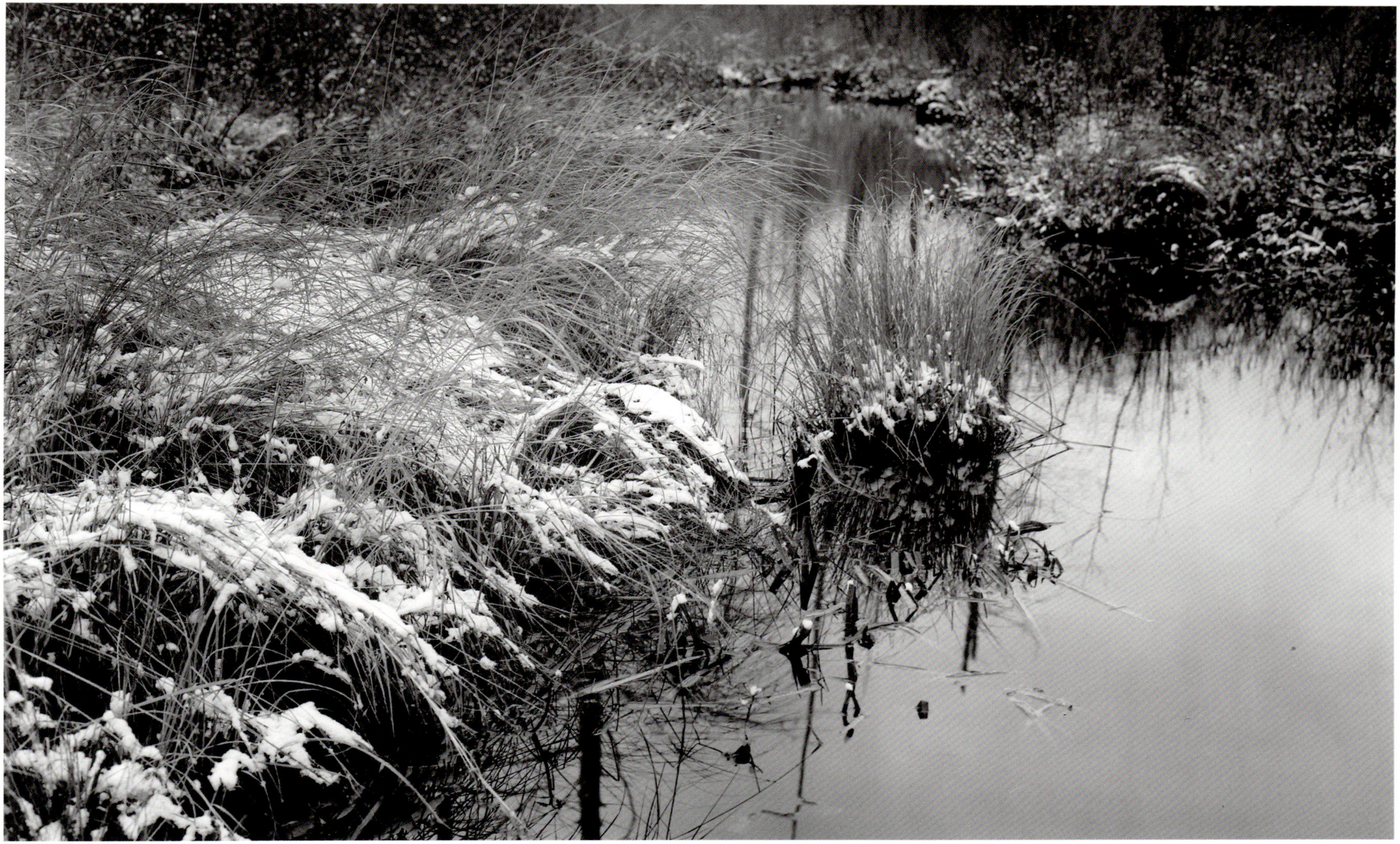

PLATE 36

CEDAR AND SNOW NEAR PORT SYDNEY 1994

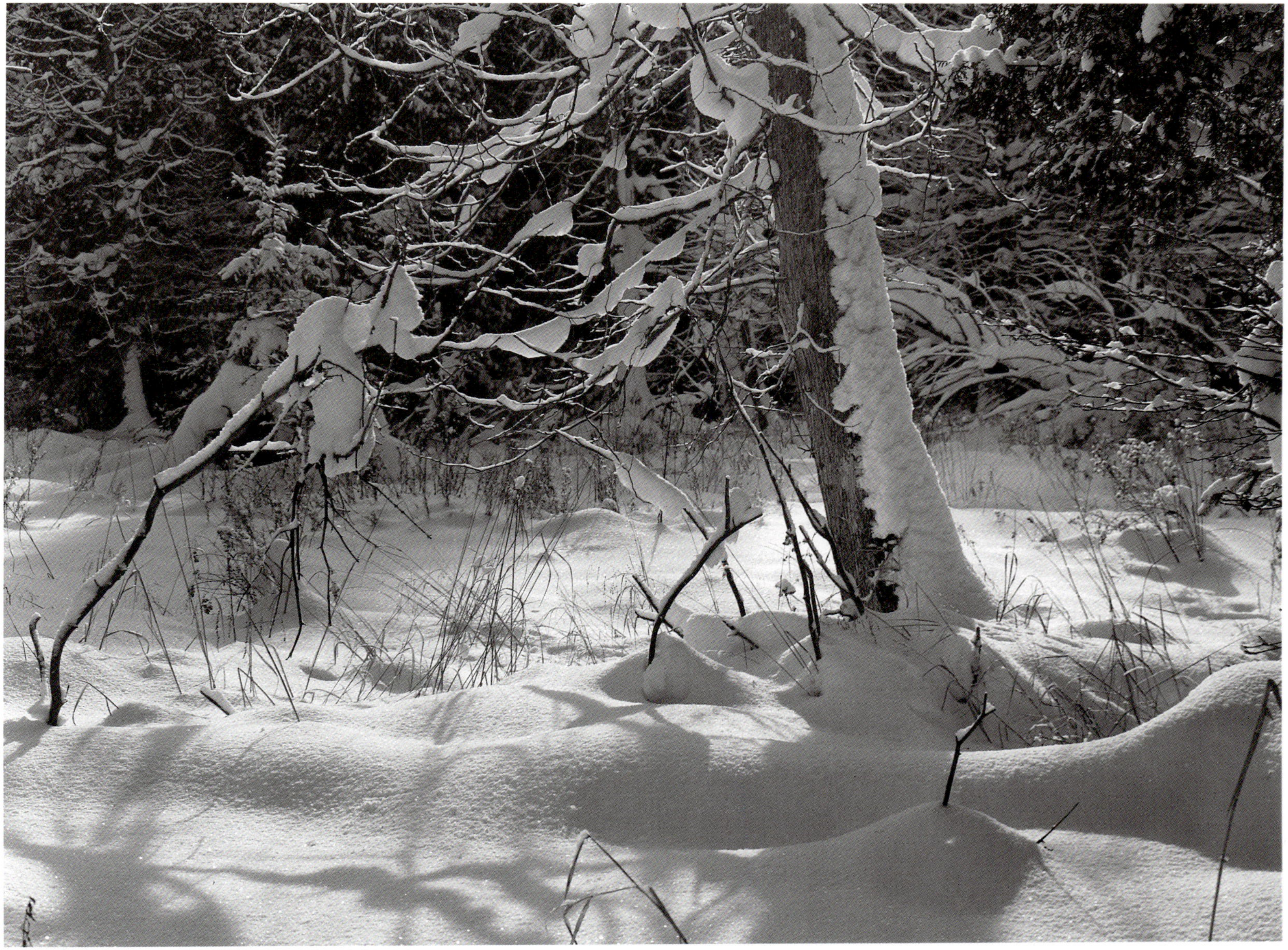

PLATE 37

SNOW MOUNDS NEAR BRACEBRIDGE 1994

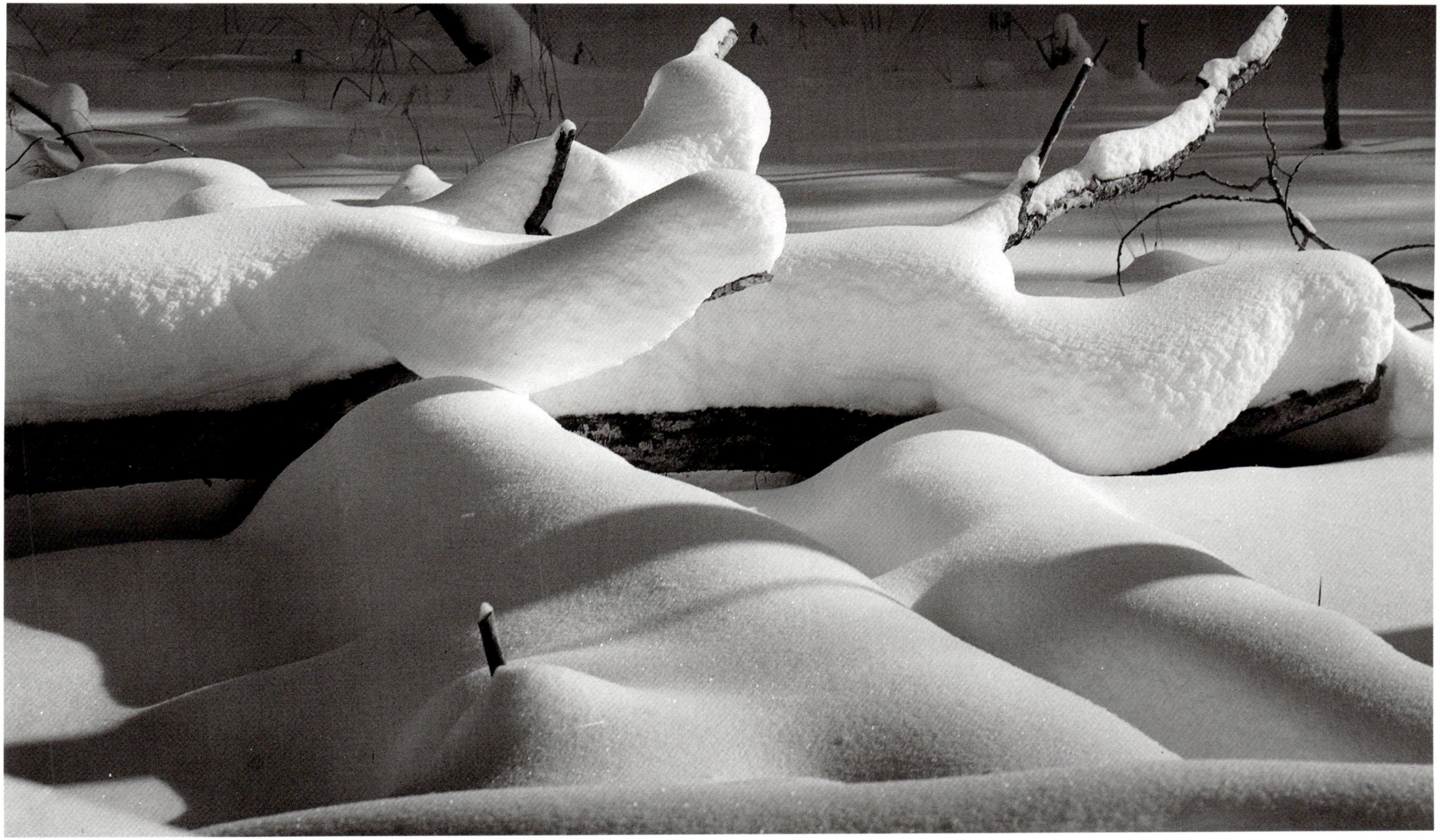

PLATE 38

HEMLOCKS AND SNOW, WILSON FALLS TRAIL 1993

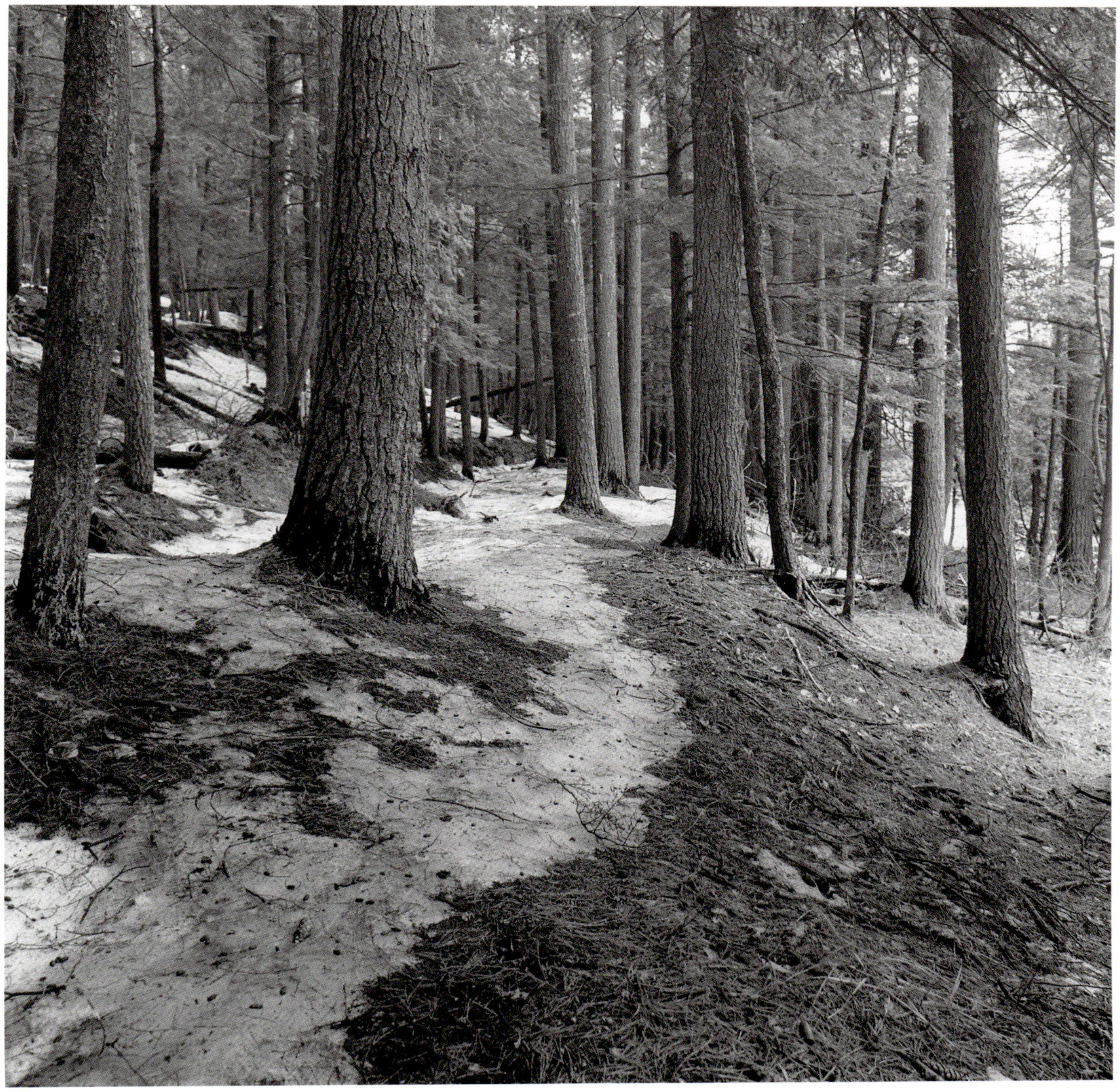

PLATE 39

SPRING THAW, HIGH FALLS AREA 1993

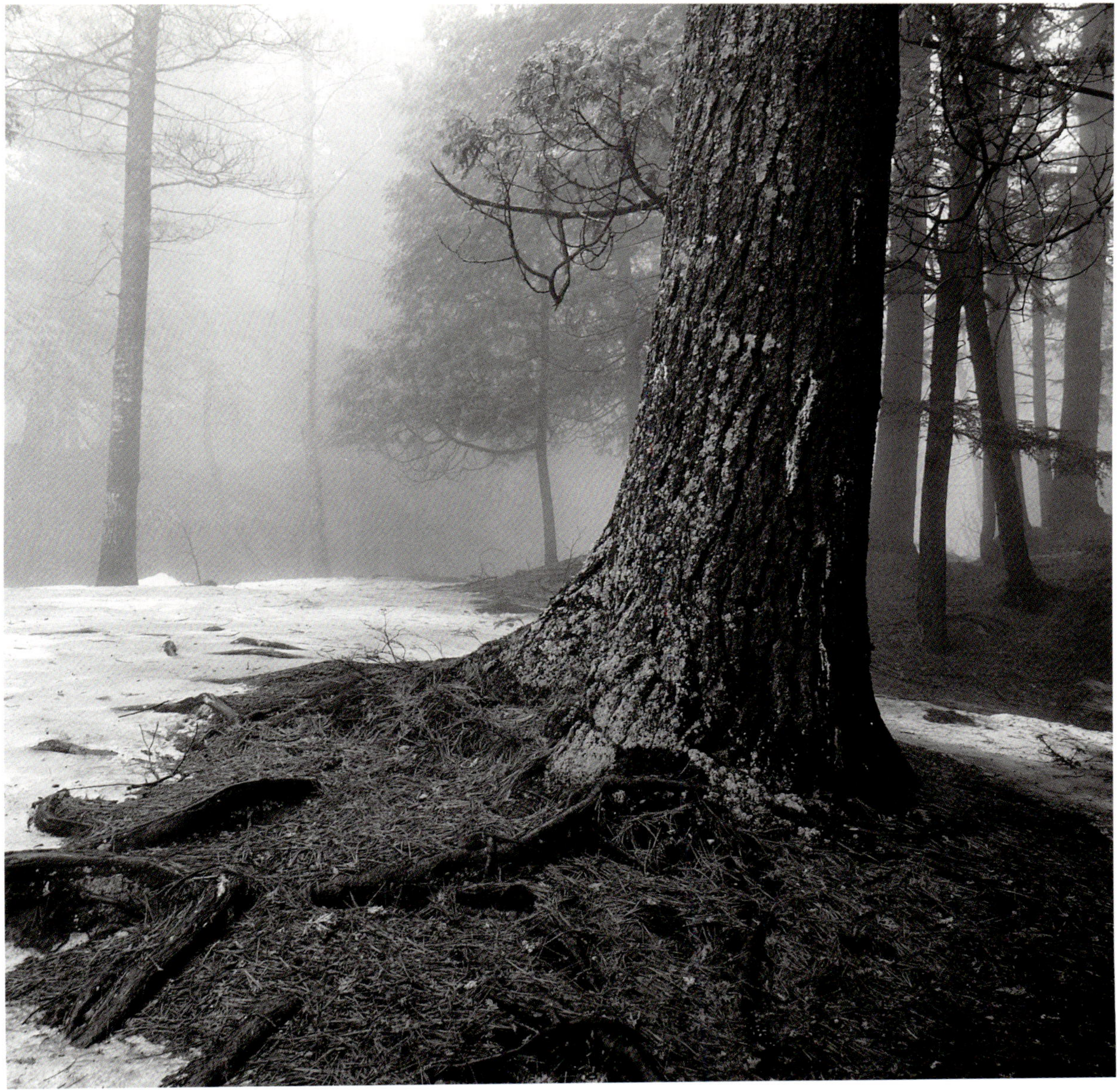

PLATE 40

Technical Information

During the past three years or so, I have dedicated my spare time to black-and-white photography, searching out new areas in which to experiment and in which to test various materials. For film, I have settled on Kodak Tmax 100 and 400.

The negatives for the prints that appear in this book were measured with a Pentax one-degree spot meter, for ultimate shadow detail when printed on a paper grade of 2.5. Shadow detail was exposed two stops below eighteen-percent grey, or zone three. Film speed was tested and adjusted to this zone. Textured highlights were measured for zone seven, two stops above eighteen-percent grey, and adjusted by testing with various developing times, again for paper grade 2.5 contrast.

My experiments with different types of photographic papers led me to Kodak Polymax RC paper, for its brightness and extended contrast.

Most of the photographs in this book were exposed on a Linhof Technika V-4x5 camera. Lenses were Rodenstock and Schneider 65mm to 350mm. Some of the other images were photographed with a 6x6 Hasselblad, and some of the most recent ones with a Horseman 4x5 Field Art camera, which impressed me with its weight and flexibility.

Testing by the trial-and-error method, while keeping strict notes on exposures, times and temperatures, proved a wonderful learning experience.

PLATE	FORMAT	CAMERA	LENS	FILM	FILTER	DEV
FRTSPC	4X5	LINHOF	135MM	T-400	-	N+2
1	4X5	LINHOF	75MM	T-100	12Y	N
2	6X6	HASSELBLAD	50MM	T-100	12Y	N
3	4X5	LINHOF	75MM	T-100	12Y	N
4	6X6	HASSELBLAD SW	38MM	T-400	-	N
5	4X5	LINHOF	75MM	T-100	12Y	N
6	4X5	LINHOF	75MM	T-100	12Y	N
7	6X6	HASSELBLAD SW	38MM	T-400	-	N+2
8	4X5	LINHOF	135MM	T-100	12Y	N
9	4X5	LINHOF	350MM	T-100	-	N+1
10	4X5	LINHOF	350MM	T-100	-	N+1
11	4X5	LINHOF	350MM	T-400	-	N+1
12	4X5	LINHOF	350MM	T-400	-	N+1
13	4X5	LINHOF	350MM	T-400	-	N+1
14	4X5	LINHOF	75MM	T-100	23	N
15	4X5	LINHOF	100MM	T-100	12Y	N
16	4X5	LINHOF	75MM	T-100	12Y	N
17	4X5	LINHOF	75MM	T-100	12Y	N
18	4X5	LINHOF	75MM	T-100	12Y	N
19	4X5	LINHOF	350MM	T-100	-	N+2
20	4X5	LINHOF	75MM	T-100	12Y	N
21	4X5	LINHOF	75MM	T-100	-	N+1
22	4X5	LINHOF	75MM	T-100	-	N+1
23	4X5	LINHOF	100MM	T-100	12Y	N
24	4X5	LINHOF	135MM	T-100	-	N+1
25	4X5	LINHOF	210MM	T-100	12Y	N
26	4X5	LINHOF	75MM	T-100	23	N
27	4X5	LINHOF	75MM	T-100	-	N
28	4X5	LINHOF	75MM	T-100	12Y	N
29	4X5	LINHOF	135MM	T-100	-	N
30	4X5	LINHOF	135MM	T-100	12Y	N+1
31	4X5	LINHOF	75MM	T-100	-	N+1
32	6X6	HASSELBLAD	50MM	T-400	-	N+2
33	4X5	LINHOF	135MM	T-400	12Y	N
34	4X5	LINHOF	135MM	T-400	-	N
35	4X5	HORSEMAN FA	135MM	T-400	12Y	N
36	4X5	HORSEMAN FA	135MM	T-400	12Y	N
37	4X5	HORSEMAN FA	135MM	T-400	12Y	N
38	4X5	HORSEMAN FA	135MM	T-400	12Y	N
39	6X6	HASSELBLAD	50MM	T-100	12Y	N
40	6X6	HASSELBLAD	50MM	T-400	-	N+1

Acknowledgments

I would like to thank my friend John Aird, his wife, Jane, and their family for their warmth and support over the past years. I would also like to thank John for his kind words in the foreword of this book.

Thanks to Gary Biggs; I hope I didn't bore you too much during our brainstorming sessions about Kodak products. I also acknowledge Kodak for their fine materials and informative support.

Thanks to John Denison, who started me off with his idea for a black-and-white fine art book.

Thanks to my wife, Janice, for her support and motivation, for listening, and for just putting up with me.

And thanks to my mentor, Al Gilbert, the master of masters in photography, who, in my earlier years of portrait photography, guided me with his friendship and his display of professional perfectionism.